A BASIC TREATISE ON THE COLOR SYSTEM OF ALBERT H. MUNSELL

A GRAMMAR OF COLOR

EDITED AND WITH AN INTRODUCTION BY FABER BIRREN

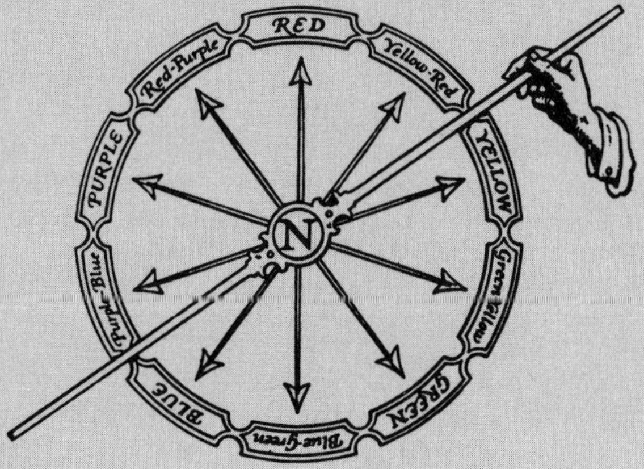

VAN NOSTRAND REINHOLD COMPANY
New York Cincinnati London Toronto Melbourne

Van Nostrand Reinhold Company Regional Offices: New York, Cincinnati, Chicago, Millbrae, Dallas
Van Nostrand Reinhold Company International Offices: London, Toronto, Melbourne

 Library of Congress Catalog Card Number 69-15896
ISBN 0-442-11343-9

Published by Van Nostrand Reinhold Company
450 West 33rd Street, New York, N.Y. 10001

3 5 7 9 11 13 15 16 14 12 10 8 6 4 2

Contents

Introduction

Albert H. Munsell was the greatest of American colorists, and his System is more widely used than any other in existence today. His conception of color harmony, of color organization and notation were original with him and mark one of the great contributions to the art and science of color order.

Yet while the Munsell System has gained world eminence as a method of color identification, Munsell's theories and principles of beauty and harmony have been less known and appreciated. One purpose of this book is to correct this by drawing new attention to Munsell's views on ideal balance as achieved through unique arrangements and combinations of color.

A few decades ago, during the twenties and thirties, Munsell was highly regarded in the field of art education. Available were Munsell papers, crayons, watercolors, plus various charts and supplies. His book *A Color Notation* was widely used. (It is still in print in an eleventh edition.) However, as expression with art and color sought new freedom, as it abandoned formal training and cast off all restrictions, Munsell's principles (as well as those of most other theorists) were set aside.

Now the abstractionists and action painters have themselves gone into decline. With new attention paid to forms of art that deal with optical and perceptional factors in color, a vital need for training, understanding, and control has returned to favor. And the contribution of Albert H. Munsell became significant and important once again.

This present book, *A Grammar of Color*, is based largely on a magnificent volume of the same title issued in 1921 by the Strathmore Paper Company. The introduction was by Munsell himself. An explanatory text, with striking charts and diagrams, was the work of T. M. Cleland, one of America's foremost graphic artists. This material has been edited and reprinted, with new chapters added which interpret, in modern terms, Munsell's concepts of beauty and the universal quality of his color order system and method of notation.

Munsell's work is in every way classical. It has developed permanent values in the art and science of color. This book, in its small way, hopes to pay further tribute to a great American and to add further renown to his fame.

FABER BIRREN

Albert H. Munsell (1858-1918)

According to notes presented by his son in *A Color Notation*, Albert H. Munsell was born in Boston in 1858. "He grew up with a great interest in out-of-door sports, but most of all he loved to spend his time on the ships and tugs of the waterfront." Living in comfortable circumstances, he took an early interest in color and art and after graduation from high school attended the Massachusetts Normal Art School.

Awarded a fellowship for foreign study, Munsell went to Paris and studied at the Acedémie Julian, a highly successful art school attended by many Americans. Here he did creditable work, took examinations for the Beaux Arts and won second prize for his painting "The Ascension of Elijah." Being scholar as well as artist, Munsell had both talent and application. In Paris he won a Catherine de'Medici scholarship and had the advantage of an extra year of study in Rome.

On his return to America he was appointed lecturer in color composition and artistic anatomy at the Normal Art School in Boston, a position he held until 1915.

It was before the turn of the twentieth century, however, that Munsell devoted himself more to color and less to painting. In 1879, at the age of twenty-one, he had read *Modern Chromatics* by Ogden Rood, professor at Columbia

College in New York. Rood's book became a masterpiece in the field of physiological optics, ran into many editions and was considered a "bible" on color by the Neo-Impressionists of France.

In 1898 Munsell created his famous Color Tree or Sphere. As an aid to teaching he visualized and devised a three-dimensional solid within which colors of infinite variety could be neatly arranged and identified by notation. And after this he perfected his system still further, saw it flourish, gain recognition by scientists, and acceptance by educators. Soon elaborate books of standards, special color charts, and art materials were being produced and distributed throughout the western world. Meanwhile he lectured extensively in America and had his achievements justly honored.

In 1914 Munsell addressed audiences in London, Paris, and Berlin. Now his health began to fail. Four years later, in 1918, he passed away at the age of sixty.

No color system, anywhere in the world, is more famous than that of Albert H. Munsell. In 1918 a number of his friends formed the Munsell Color Company to carry on his work. The Company, located in Baltimore, continues to offer color standards and has become a practical clearing house concerned with problems of "standardization, nomenclature, and specification of color."

The Munsell system itself has been improved, extended, and given technical perfection by a group of scientists. For its method of color notation it is used throughout America as a national system. It also has acceptance abroad, particularly in Great Britain and Japan.

The Color Sphere

(The three brief chapters that follow on The Color Sphere, Balance of Color, and Unbalance of Color are the words of Albert H. Munsell as written for *A Grammar of Color*. Munsell died in 1918 and the book was published in 1921, so the text appeared posthumously. Out of consideration for the reader quotation marks have been avoided and a slight amount of editing done to simplify meanings.) Munsell speaks:

A clear mental image of color relations must underlie any intelligent grouping of its hues in the best degrees of strength and light. This image is best produced by using a sphere to represent the world of color. (See Figure 1.) With white at the North pole and black at the South pole; and its axis between these points a measured scale of grays, we have a decimal neutral scale which painters call Value. The middle point of this axis must be a middle gray and a plane passing through to the equator must contain colors of middle value. If therefore the equator be spread with a color circle of Red, Yellow, Green, Blue, Purple; and the half-way points by their mixtures in Yellow-Red, Green-Yellow, Blue-Green, Purple-Blue, Red-Purple, we have the equator as a decimal scale of hues merging gradually from one to the next and returning upon itself at Red.

Each of these hues is supposed to grow lighter until it merges into the North pole at white, and darker similarly to black, and these are called the values (light) of color. They may also be imagined as passing inward until they disappear in the gray axis. Should there be still stronger colors they will continue upon the same radii outside the sphere. These we call the Chromas (strength) of color. In this way every point inside of the sphere and some upon the outside are arranged in three scales as follows: a vertical scale of light values, a horizontal scale of Chromas, and a circular scale of Hues; and

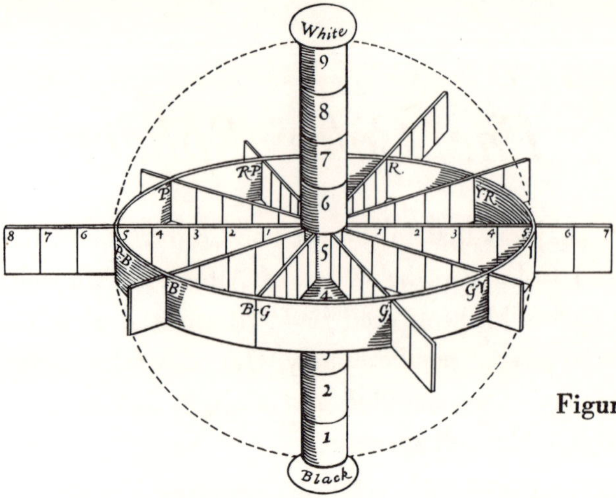

Figure 1.

since these are all in decimal divisions it becomes easy to make it a permanent mental image in which to see all color relations.

Naturally every point in these three scales has its defined number, just as a solid object has its three dimensions; and to write them as a symbol of that color, thus doing away with the foolish misleading names which are prevalent, we have only to image the three angles of a triangle occupied with the three parts of that symbol — the left hand angle by the Hue initial (Red, Yellow-Red, etc.); the upper angle by a number describing its value in the scale of light; and the right hand angle by a similar number describing its Chroma in the scale from the axis outward. Thus, vermilion has for its symbol R 5/10.

This may seem revolutionary to anyone who has heard no end of fanciful names which fail to describe colors; but each symbol accurately describes the color in its dimensions of Hue, Value, and Chroma.

This has all been worked out in permanent color in the complete Munsell Color System and each step bears its permanent symbol. There can be no new color discovered for which a place and symbol is not waiting. With this system in mind it is as easy to understand color relations as to understand musical relations on the written score. Indeed it furnishes the written score which is described in the handbook *A Color Notation.*

Balance of Color

Munsell continues to speak:

The sense of comfort is the outcome of balance, while marked unbalance immediately urges a corrective. That this approximate balance is desirable may be shown by reference to our behavior, as to temperatures, quality of smoothness and roughness, degrees of light and dark, proportion of work and rest. One special application of this quality is balance which underlies beautiful color. The use of strongest colors only fatigues the eyes, which is also true of the weakest colors. In a broad way we may say that color balances on middle gray. Thus a moderate amount of extremely strong color may be balanced by a right quantity of grayer color; and a brilliant point of strong red will balance a larger field of the grayest blue-green.

Thus AREA is another quality in color composition, which aids in the balance of Hues, Values, and Chromas. Examples of this are all about us. The circus wagon and poster, although they yell successfully for our momentary attention, soon become so painful to the vision that we turn from them. Other examples are magazine covers and theatrical billboards. These are all cases where color is used only to excite the eye but not for its permanent pleasure.

The large truth is that general color balances approximately upon middle gray. Although the colors may differ greatly, yet their total effect is balance.

Let us take a point upon the color sphere such as R 5/5. (See Figure 2.) There are three distinct color paths for which this becomes the center. First a vertical path which extends from black through red to white; and in a decimal system is divided into ten equal steps. Equal departures either way from middle red must balance, such as R 7 with R 3, R 8 with R 2, R 6 with R 4, while the strength may be used so as to require equal or unequal areas of each balanced pair — the general law being that the stronger the color we wish to

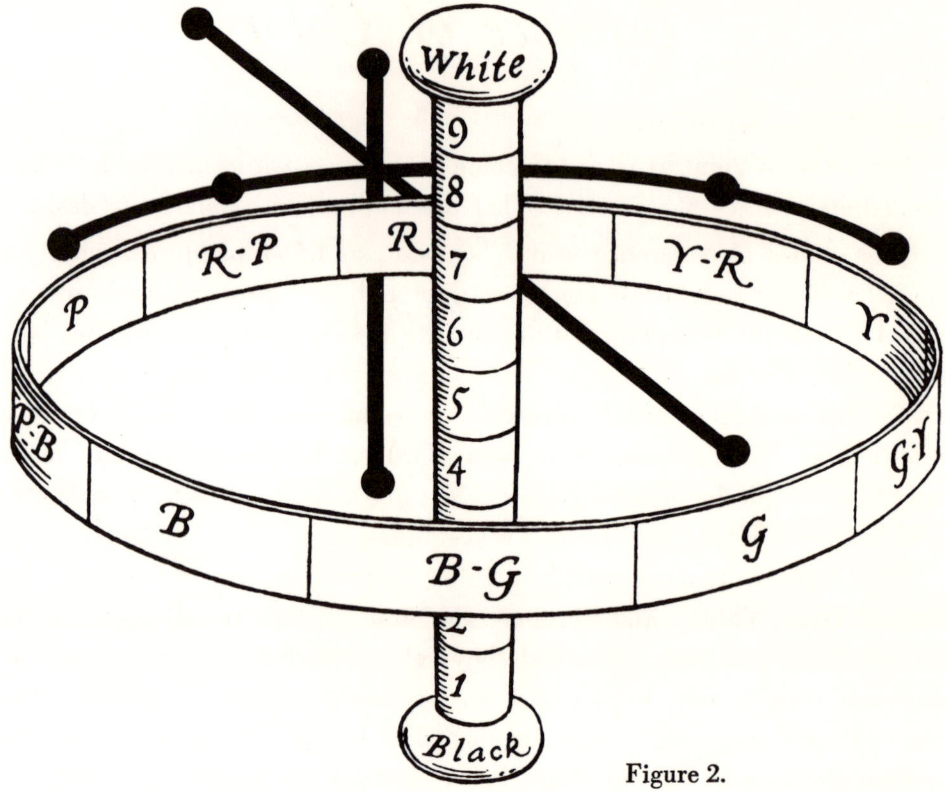

Figure 2.

employ, the smaller must be its area, while the larger the area, the grayer the Chroma. Thus R 7/6 balances R 3/3 in the proportion of nine parts of the lighter red to forty-two parts of the darker red. In other words, these symbols will balance colors *inversely* as the product of their factors. This opens up a great field of *area* in the use of reds, where balance may be restored by changes in the factors of Value and Chroma. Thus the lighter red (R 7/6), which we will call 42, balances the darker red 3/3, which we will call 9, by giving 42 parts of the darker (weaker) red to 9 parts of lighter (stronger) red.

A second path through Middle Red follows the equator of the sphere and again we may balance the Hues once or twice removed, as, for instance, RP and YR or P and Y. These are called the neighbors of Red, popularly known as its shades. Instead of neighbors we may select the exact opposite of Red, that is, Blue-Green, which is known as its complement, using equal areas if the colors are of equal strength or increasing the area of the weaker color. This second path does not depart from the level of the equator and therefore all the colors named are of a single Value without contrast of light and dark.

A more interesting path is the third, which may be passed through Middle Red, being neither vertical nor horizontal; but inclined so that if it passes upward out of Red toward lighter Purple it will pass downward from Red into darker Yellow.

These three examples must suffice as a brief introduction to almost endless examples of color series and color intervals that are orderly and harmonious to the eye.

Unbalance of Color

Munsell concludes:

That any long duration of unbalance, either mental, physical or spiritual is an aggravated form of disease may be easily shown. Yet short periods of unbalance are very stimulating in the effort which they produce to regain balance. We see this in the introduction of discords in music. In contortions of the body. In intentional inversions of thought. This also shows in the seasoning of our food. Too sweet, too salt, too sour. It even shows in our criticism of pictures. We say, too light, too dark, too hot, too cold, too weak, too strong, and the effort of the accomplished artist is to overcome these forms of unbalance.

The introduction of a color scheme of a certain moment of unbalance is called harsh color, it leads to its correction by what we call harmonious color (really balance); and the contrast enhances the latter; so that to overcome monotony, we should be able to use unbalance wisely at times, in order that the general balance may be the more evident. This is sometimes done in the picture gallery by means of a so-called "gallery of horrors;" — in music by a sudden discord; in behavior by an unexpected rudeness; — all illustrations of the value of the contrast between harmony and discord; and this quality of contrast is proportioned to the use of color. If it is to serve as the background of the picture, the color must be quiet. If it is to be the makeup of the pictures themselves there must be strong oscillations in the contrasts of light and dark (Value), of hot and cold (Hue), of weak and strong (Chroma).

As in the case of advertising color, especially in the open air, the very strongest contrasts and even strident relations are admissible. Any attempt in this sketch to encompass this broad question of color harmony would be impossible.

(Yet in chapters that follow, written by T. M. Cleland and Faber Birren, considerable attention and detail will be paid to Munsell's unique and worthy principles of balance and unbalance in the use of color.)

The Ascension of Elijah, Albert H. Munsell, ca. 1885.

The Munsell System

This rather lengthy chapter was written — and illustrated — for *A Grammar of Color* (1921) by T. M. Cleland, one of America's foremost graphic artists. Cleland had a scholarly mind and greatly admired Munsell. What he had to say with words and what he portrayed with his drawings and charts are rare tribute indeed. The reader will note that he thoroughly discusses the organization and makeup of the Munsell System, as well as analyzing Munsell's principles of Complementary Colors, Balance, and Color Combinations. These principles of harmony will be dealt with in greater detail in the last part of this book — and pictured in a series of color plates. Modern applications of Munsell's notations and color identification methods will also be elaborated upon. Cleland speaks at follows:

In the three introductory chapters written by Professor Munsell will be found a brief compendium of his theories upon the dimensions of color and color relations, which, though generally scientific in form, is stated with such admirable simplicity and absence of scientific verbiage that it merits careful study by all practical workers who would understand the basic idea upon which the matter of this book is built. It has been thought wise, however, to augment this with a practical description with illustrations of the cardinal principles of the Munsell System, more especially with a view to its actual use in what has come to be generally known as the Graphic Arts. In so doing, there must necessarily occur a reiteration of much that appears in Professor Munsell's introductory statements, but its being expressed in different form may tend to assist the practical reader toward a clearer comprehension.

The first essential to the application of the Munsell System is a clear understanding of the three dimensions of color, and once having grasped the simple logic of these, the practical advantages of the System will be manifest. The reader should be warned at the outset against that fear of scientific perplexity which is ever present in the lay mind. The three dimensions of color are not involved in the mysteries of higher mathematics. There is nothing about them which should not be as readily comprehended by the average reader as the three dimensions of a box, or any other form which can be felt or seen. We have been unaccustomed to regarding color with any sense of order, and it is this fact, rather than any complexity inherent in the idea itself, which will be the source of whatever difficulty may be encountered by the reader who faces this conception of color for the first time.

In Color Plate I will be found a colored diagram, accompanied by an explanation, which has been made especially to present the three dimensions concretely and to avoid the abstractions of written explanation. The idea of the three dimensions of color can be even more simply, though less completely, expressed as in Figure 3.

With these three simple directions of measurement well in mind, and by reference to Color Plate I mentioned above, where actual colors are printed, there need be little confusion for even the least scientific mind in compre-

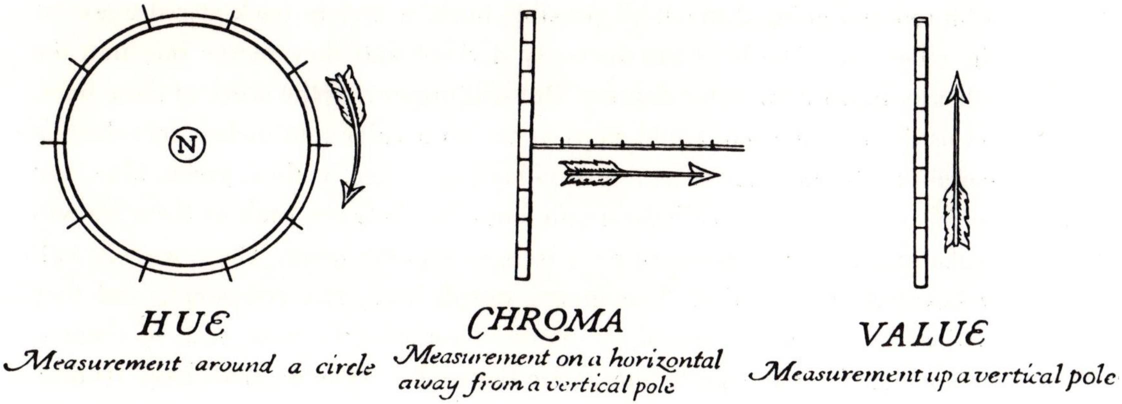

Figure 3.

hending what is meant by color "measurement." In considering further the qualities of color, which are expressed by these three dimensions known as Hue, Value, and Chroma, we will take each one of them separately in the order in which they are written, trusting that having done so we may pass to the subject of color balance or harmony and its application to every-day practice equipped with a clear understanding of how it may be measured and noted.

HUE

This first dimension is defined by Professor Munsell as "The quality by which we distinguish one color from another, as a red from a yellow, a green, a blue, or a purple," but this dimension does not tell us whether the color is dark or light, or strong or weak. It merely refers to some point in the spectrum of all colors, such as we have seen in the refraction of sunlight through a prism. Let us suppose now that we had such a spectrum cast by a prism, or a section taken out of a rainbow. We know it to be a scientific fact that it contains all possible hues, merging by indistinguishable degrees one into the other, but always in a fixed order. Now let us imagine that we have such a spectrum fixed or printed on a band of paper, and that it begins at one end with red and going through all possible hues, it arrives back at red again at the other end. The hues are unevenly divided and they merge one into the other by indistinguishable degrees. But still preserving the order of these hues, let us divide them into equal steps as we do a ruler into inches, by selecting certain colors familiar to us in every-day use — red, yellow, green, blue, and purple. These we will call the simple hues, but between each of them we will make another division where each merges into the other. These we will call yellow-red, green-yellow, blue-green, purple-blue, and red-purple and they will be known as compound or intermediate hues, because each of them is compounded of two simple hues. (See Figure 4.) Thus we shall have 10 divi-

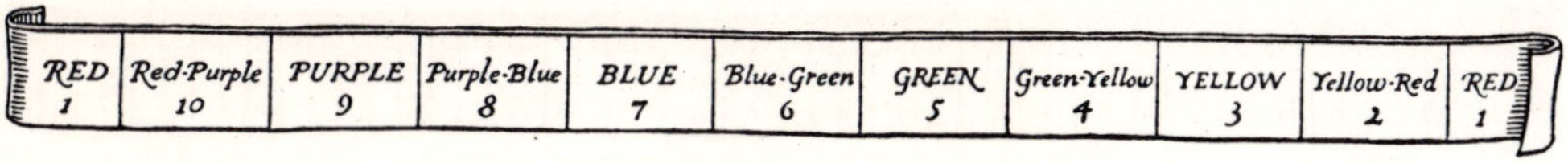

Figure 4.

sions upon our band. The reason for this number of divisions will be understood when we come to consider the question of color balance. It presents a sufficient variety of hues for purposes of demonstration and for most practical uses. Now if we bend this band around into a circular hoop, so that the red at one end meets and laps the red at the other end, we have a perfect scale of hue in the circular form in which we shall always consider it. (See Figure 5.) So it is that when we state the first dimension of a color we are merely referring to its position on this circle of hues. In writing a color formula this first di-

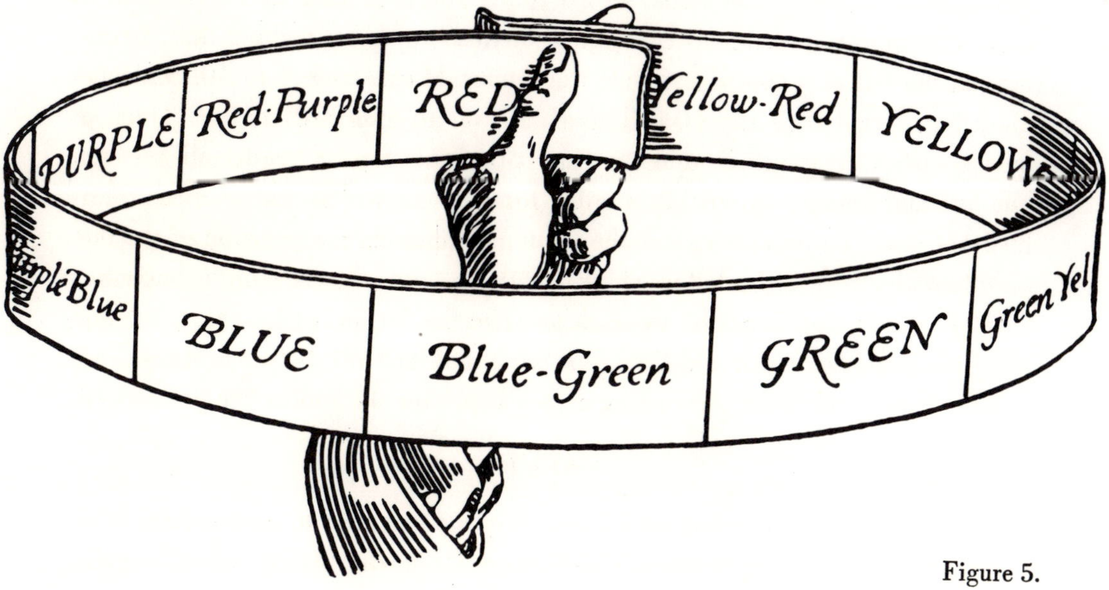

Figure 5.

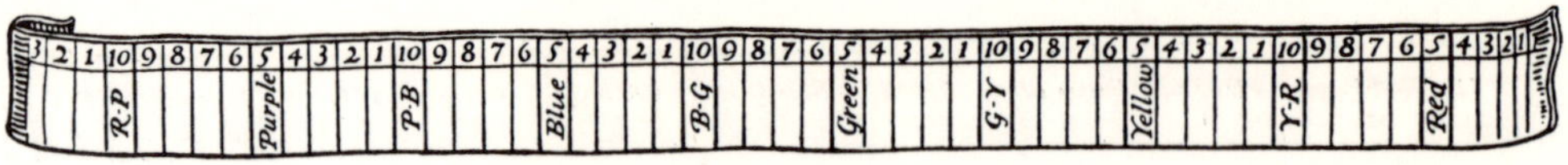

Figure 6.

mension is expressed by the initial letter of the hue — R for red, which is a simple hue, and BG for blue-green, which is an intermediate hue.

These 10 steps being a decimal number, may, of course, be infinitely subdivided and it may frequently happen that a given color does not fall exactly on any one of these 10 divisions of hue, but somewhere between two of them. Allowance has been made for this by dividing each of the steps of the simple hues into 10 further divisions. These 10 subdivisions represent about as fine a variation of hue as even a trained eye can distinguish, and it would be obviously futile, for practical purposes, to carry it further. If we uncurl our band again, in order to better see what we are doing and note these divisions upon it, they will appear in order as on Figure 6. Reading from right to left, beginning at the left of a compound hue, the numerals run from 1 to 10, 5 always marking a simple hue and 10 falling always on an intermediate hue. Thus we have a series of numerals denoting any practical step or gradation between one hue and another in writing a color formula, of which one of these intermediary hues is a part; we place the numeral, denoting the position of the hue on this scale, before the letter which stands for the nearest simple hue, thus 7R, 2Y, etc. If, for example, we wish to write the formula of a color, the hue of which is neither red nor yellow-red, but about halfway between the two, we would write it 7R or 8R, according as it was nearer the red or the yellow-red.

VALUE

This is the second dimension and is possibly the simplest to understand. It is, according to Professor Munsell's definition, "The quality by which we distinguish a light color from a dark one." We noted that the first dimension did not tell us whether a color was light or dark. It told us, for example, that it was red and not green, but we know that there may be light red and dark red, and it is the function of this dimension of Value to tell us how light or how dark a given color may be. For this purpose we shall need a scale of value, which we may conceive as a vertical pole, or axis to our circle of hues, black at the lower end, representing total absence of light, and white at the top, representing pure light, and between these a number of divisions of gray, regularly graded between black and white. This gradation could also be infinite. Since pure black is unattainable, we will call that 0 and begin our scale with the darkest gray as 1, numbering the steps up to 9, which is the lightest gray. Pure white, which is also unattainable, we will call 10. In the practical use of the scale of value, therefore, we shall have but 9 steps and the middle one of these will be 5 — what is referred to as middle value. This scale of value, or neutral pole, is well represented on Color Plate I already referred to, where it is shown with the actual gradations printed. (Also see Figure 7.) These steps of value have been scientifically measured and registered by means of an instrument known as a photometer. In writing a color formula we express this dimension of value by a numeral, which denotes at what step upon the scale of value this color falls. This numeral is written above a line, as B 6/ for example, by which we mean that this particular blue, regardless of its other qualities is as light or as dark as the 6th step upon the scale of value. A color such as is commonly called "*maroon*" is an example of a red which is *low in value*, because it is dark, and what is called "*pink*" is a red which is *high in value* because it is light.

Now having familiarized ourselves with these two dimensions, and understanding what qualities of a color they express, we may proceed to consider the third dimension, without which our description of any given color is incomplete.

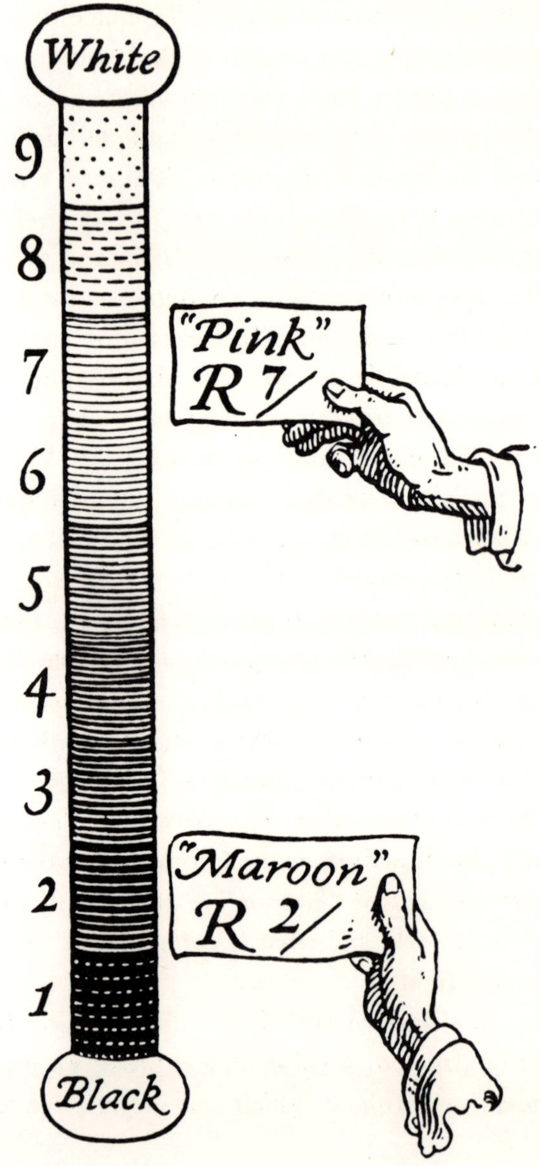

Figure 7.

CHROMA

When we have stated that the color is blue or yellow or green and that it is dark or light, we have indicated two of its important qualities — its Hue and its Value, but we have by no means described it completely. We may say of an emerald that it is green and that it is light, but we can say that certain grapes are green and also light, and yet there is a decided difference between their respective colors, if we place them side by side. Both may be green and of the same value of light, but the emerald is *strong* in color and the grape is *weak* in color or *grayer*. It is this difference which is measured on the dimension of Chroma. The scale of value has been referred to in the convenient and easily understood form of a vertical pole (Figure 7), which represents a neutral axis to all the circle of hues and is, itself, of no color, but is pure gray. Around this pole we may place our band representing the scale of hue and then if we imagine any one of these hues on the circumference of the band to grow inward toward the gray pole in the center, growing grayer or weaker in color strength until it reaches this center pole and loses its color entirely, we have grasped the idea of the dimension known as chroma. (See Figure 8.) By dividing this into regular measured steps, we have a scale upon which the

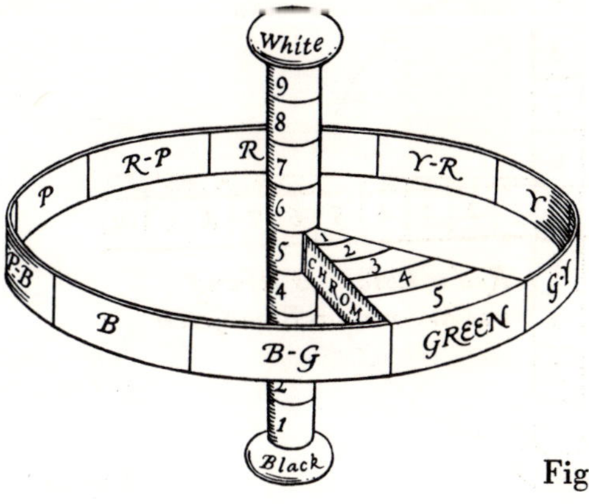

Figure 8.

strength of color may be measured. This is clearly illustrated on Color Plate I, where several steps of yellow are shown printed on the scale of chroma. This dimension of chroma is written in a color formula by means of a numeral *below* a line, which denotes the step upon the chroma scale at which it falls, thus /5, /8, /9, etc. Plate I also includes at upper right and lower left Munsell's simple and compound hues as they appear at full chroma.

Needless to say, all of the hues may be thus measured on this dimension at right angles to the vertical pole and grading from gray, step by step away from the pole to greater and greater strength of color.

Professor Munsell has devoted a part of his introductory chapters to a description of what he calls "The Color Sphere." (Figure 1.) This is a general form which aids the orderly consideration of color and within which all color balances, as will be shown later; but in the actual measurement of pigment colors, such as we use in printing or painting, all of the paths of chroma would not be of the same length nor would they all be comprised within a sphere. Certain of them would extend to points outside of it. Nor would all of

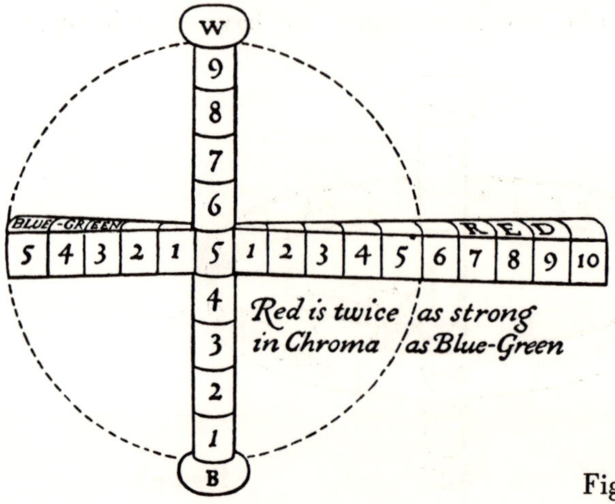

Figure 9.

the paths of chroma reach their greatest length at the equator of the sphere, that is, the level of middle value. There are two reasons governing this which it is important to understand: first, *colors differ by nature in their chroma strength, some being much more powerful than others.* The strongest red pigment used, for example, is twice as powerful as the strongest blue-green pigment and will require a correspondingly greater number of steps on a longer path to reach gray. The chroma path of red is the longest and extends far outside the sphere, being ten measured steps from the neutral pole, while blue-green is the shortest, being only five steps. (See Figure 9.) The sphere is limited in size to this shortest axis for reasons which will appear when we take up the question of balance or harmony of color. The second reason is: *That all colors do not reach their maximum chroma strength at the same level of value.* It can be readily comprehended, for example, that the strongest yellow pigment is by nature much lighter, or higher in value, than the strongest blue pigment and, therefore, that the complete chroma paths of these two colors will each touch the neutral pole at different levels. (See Figure 10 and Color Plate I.)

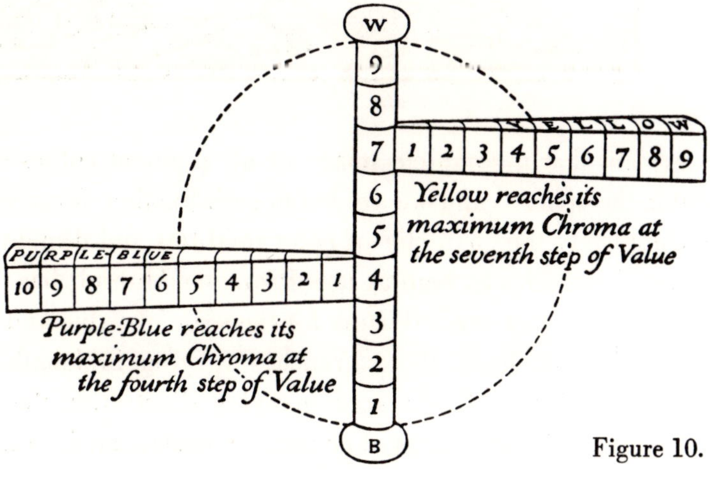

Figure 10.

Figure 11.

Thus it is evident that a complete image of all pigment colors cannot be comprised within the sphere; and we are led to seek another form which will convey more completely the character of color qualities and dimensions governing the range of pigments in regular use. Professor Munsell has conceived this as a "Color Tree" with a vertical trunk for the scale of value and branches representing the different hues, these branches varying in length with the chroma strength of each hue. In the accompanying illustration the leaves of the tree represent the measured steps of chroma on each branch (Figure 11).

Upon the scale of chroma the number of steps is limited only by the strength of pigments. The strongest yellow pigment in dry form, for example, will reach nine steps away from the neutral pole; but certain dyes on silk, or even printing inks and some unreliable pigments, may go one or more steps beyond this. As new and more powerful pigments may be discovered, they will add further steps to the scale of chroma. (Color Plate II represents a complete section of the Munsell Color Tree or Sphere in which yellow is shown in all its various values and chromas. Note that the yellow hue reaches its maximum chroma at value 8. Other major hues, being deeper in quality than yellow, will naturally reach full purity at lower values. See also chapter on "The Science of Identification" at the end of this book.)

We have described each of the three dimensions by which any color may be measured, and noted how each is written in a color formula. It remains only to put these separate notations together and to write a complete color formula embodying all three dimensions. For example, we are given a certain color to measure and define and we find that upon the scale of hue it is purple-blue. Upon comparing it with the scale of value, we find it is but three steps from the bottom, and that it is only two steps away from the neutral gray pole upon the scale of chroma. A complete formula for this color would, therefore, be written PB 3/2. It is scarcely necessary to point out the practical advantages of such a system of definite measurement and notation over the vague and variable terms in general use, borrowed from the vegetable and animal kingdoms, such as plum, olive, fawn, mouse, etc., of which no two persons ever have quite the same idea.

It is hoped that the foregoing explanation of the three dimensions of color will have been sufficiently clear to convey to the reader a distinct mental image of what is meant by the terms Hue, Value, and Chroma, in order that we may proceed to the study of certain principles of order for the intelligent and harmonious use of color, which grow out of this simple and logical system.

OPPOSITE OR COMPLEMENTARY COLORS

Figure 12, displaying a circle of ten regular hues arranged in the immutable order imposed by the spectrum, and reading clockwise, beginning with red at the top, will serve, with but little explanation, to illustrate what is meant by "opposite," or the possibly more familiar word "complementary," colors. The term opposite is used preferably in the Munsell System because it is simple and is self-explanatory, as will be seen by reference to the shown diagram, where each hue on the circle will be found directly opposite to another hue. Thus a straight line drawn from red on the circle of hues through the neutral pole will pass through blue-green, its opposite or complementary color. A line from blue through the neutral pole will pass through yellow-red and so on throughout the whole circle. It should be noted that each of the simple hues, red, purple, blue, green, and yellow falls opposite a compound hue, blue-green, green-yellow, yellow-red, etc. Now two colors which are thus opposite to one another are not only farthest apart upon the diagram, but are in actual use the most strongly contrasting. It does not matter at what point we draw the line, whether it is from one of the regular hues or from a point between two hues, if it passes through the center it will fall upon the hue or intermediary hue which is its strongest contrast. This may be more readily visualized if we imagine the spindle indicated on the diagram as pivoted on the neutral pole and movable to any point on the circle. The question may be asked as to how it is determined that these colors, which fall opposite to one another on the scale of hue, are, in fact, the most strongly contrasting colors. The answer to this question will serve to demonstrate the logical foundation of the Munsell System. When any two colors are truly opposite or at the point of strongest contrast, their admixture will produce a perfectly neutral gray. Though this may be accepted as axiomatic, it can be easily proven with scientific accuracy by arranging two opposite colors on a disc in proportions relative to the

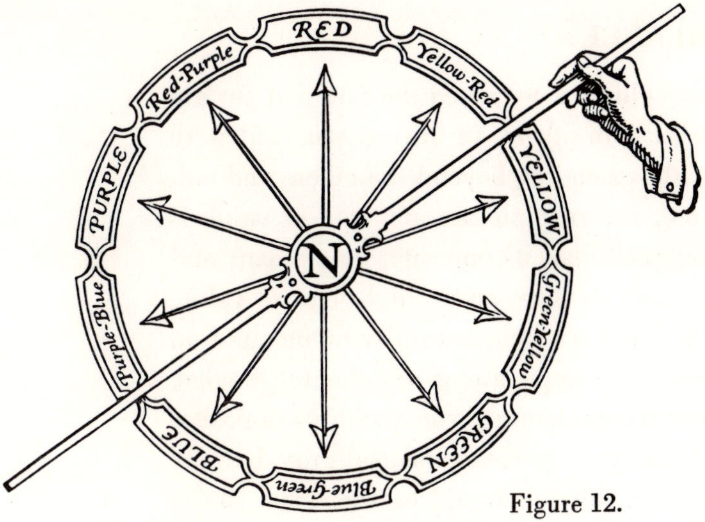

Figure 12.

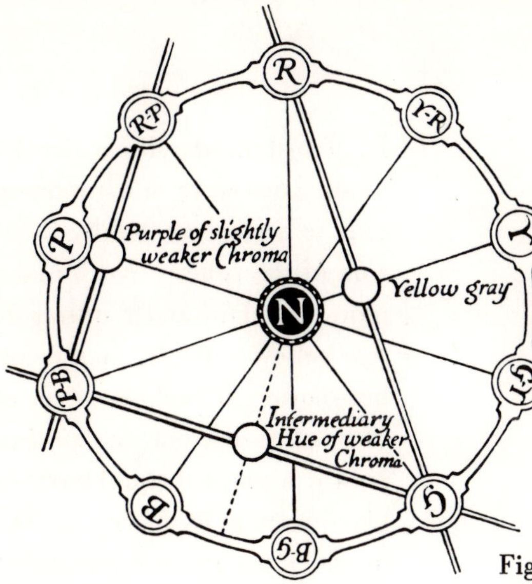

Figure 13.

chroma strength of each and revolving them with such rapidity that we cannot see them separately and they are mixed, when, if they are truly opposite, they will unite in a perfect gray. Therefore working back from this fact, the scale of hue has been so composed that those colors which thus mixed with each other do actually make gray are placed directly opposite on a line running through the neutral gray pole. Another question which may arise is what will take place if we draw a straight line between two hues which are not opposites; and what would be the result of the admixture of these. This can best be answered by the accompanying diagram, Figure 13, where three different lines have been drawn, no one of them through the neutral center. These lines, it will at once be seen, cross points which are not neutral, but nearer to one or another of the hues lying between the ones from which the lines are drawn; and the result of the admixture obtained is noted on the diagram. This will be sufficient to further demonstrate the simplicity and logic of the System and to suggest to the reader other interesting examples of it.

BALANCE

In describing the dimension known as chroma, we noted the fact that certain of the hues were much more powerful than others, in this regard, and were only to be represented by lines or paths extending beyond the others and outside of the sphere. We found that red, for example, on any step of value is more powerful and requires a longer path than its opposite, blue-green; and that yellow is longer than its opposite, purple-blue, on the high steps of value, but shorter on the lower steps of value. This brings us naturally to the question of balance of color, the *vital* question in all applications of color to practice. Now if we mixed equal parts of red at its maximum chroma with its opposite, blue-green, at its maximum, we would not get a perfectly neutral gray, but one in which the red predominated very decidedly. It would be somewhat like a tug-of-war in which there were ten men, each representing a step of chroma, on one side and only five on the other. (See Figure 14.) The resulting color would be pulled well over on to the red side, because of the fact already stated that red at its maximum chroma is so much stronger than blue-green at its maximum chroma. If, however, instead of taking equal amounts of the two colors, that is to say equal quantities of pigment or equal printed areas of each, we take what would correspond to an equal number of steps upon the scale of chroma, we find that they do balance and produce a perfectly neutral gray in which neither the one hue nor the other predominates. Let us glance

Figure 14.

Figure 15.

Figure 16.

for a moment at Figures 15 and 16 in which a bar represents the line of red and blue-green, with five steps of chroma for blue-green and ten steps of chroma for red, as is the case with these two hues at middle value. The bar rests upon a fulcrum at the neutral point and obviously it will not balance, but will fall to the red side, as in Figure 15. But if we cut off steps 6, 7, 8, 9, and 10 from the red side of the bar, it will balance upon the neutral gray, as in Figure 16. This will doubtless strike the reader as so simple and obvious that it scarcely merits statement; but it is just this simplicity which is characteristic of the Munsell System throughout, if approached from the same point of view. This, too, will explain why the diameter of our Color Sphere is limited to the shortest chroma path at middle value. It will at once be apparent that within a sphere thus limited, all opposite colors will balance because being all of equal length at each level of value no chroma path can be longer than another or outbalance it.

Thus we see how two opposite colors may be balanced by employing only equal chroma steps of each on the same level of value, that R 5/5 will balance BG 5/5, or G 5/3 will balance RP 5/3 and so on throughout all of the hues. But in practice we may wish to employ a weak chroma of one hue with a strong chroma of its opposite. In this case we cannot resort to the simple expedient of chopping off the excess strength of color on one end of the line, but must attain the desired balance by another means. If our purpose is merely to make a perfect gray, we would use a greater amount of the weaker color; but if, as in general practice, we wish to produce a balanced or harmonious color design, we would employ a larger area of the weaker color than of the stronger. If we do this in correct proportions, relative to the strength of chroma in each of the colors, we will attain balance. We may prove that we have attained balance by the fact that everything in our design, thus apportioned as to area and strength of chroma, if mixed together would produce a perfect gray. Let us suppose, for example, that we wish to employ in our design the maximum of red and blue-green at middle value. Since we are speaking of balance, a pair of scales is an apt figure with which to illustrate the point. Into the pan on one side we will put *five* blocks of red 5/10, its maximum chroma. In order to balance this we must put into the other pan *ten* blocks of the strongest blue-green, which is only 5/5. (See Figure 17.)

So we find that in order to balance two colors of unequal chroma, but of the same value, we use a larger area of the weaker chroma with a lesser area of

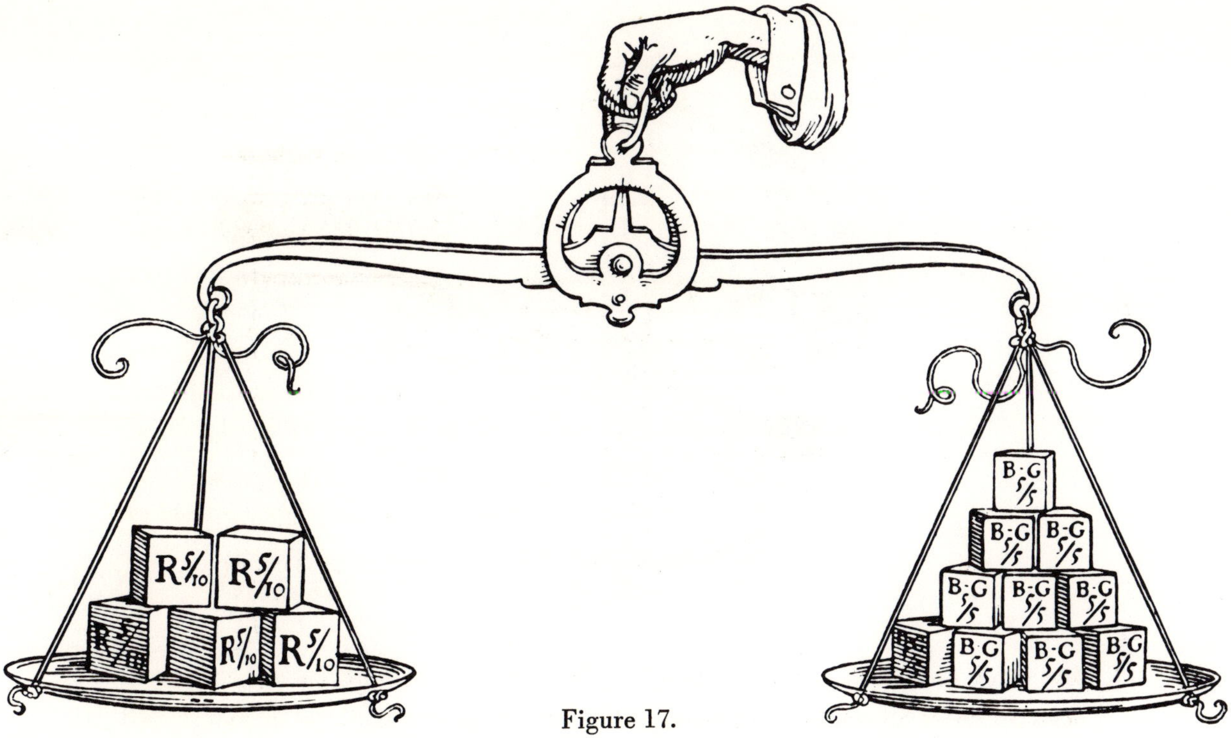

Figure 17.

the stronger, and that the proportions are simply in inverse ratio to the strength of chroma of each. That is, we use ten parts of blue-green at /5 chroma with five parts of red at /10 chroma, or let us say six parts of yellow-red 3/4 with four parts of blue 3/6, etc.

Thus far we have considered only balance of opposite hues on the same level of value; but more often than not it will occur that we wish to print a design in colors which are not only different in chroma strength but also on different levels of value, and this difference of value will also affect the question of balance and of the amount of area which each color should occupy in order to attain it. Let us assume that we wish to print a design in yellow of a high value and strong chroma, say Y 7/9, with its opposite, purple-blue, at

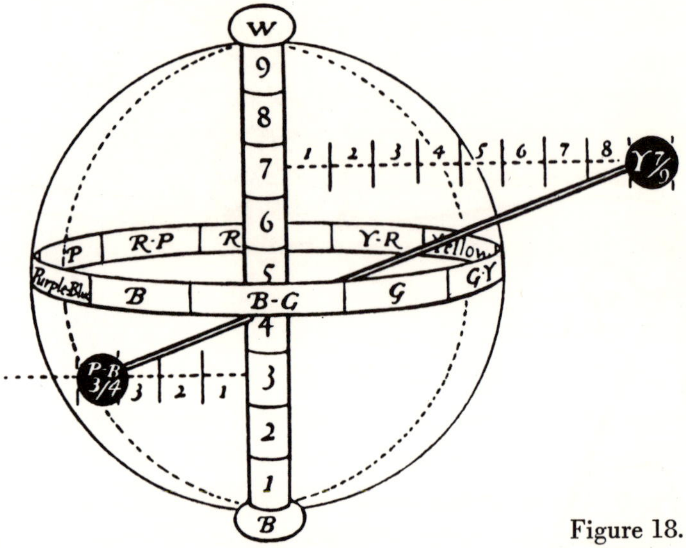

Figure 18.

low value and weak chroma, say PB 3/4. The path formed by a line drawn between these colors, passing through the neutral pole would not be horizontal in this case, since they are at different levels of value, but would appear as in Figure 18.

We now have to take the value into account in determining the amount of area of each of these two colors to be used if we are to arrive at a perfectly balanced color design; and this is done by the simple process of multiplying the chroma by the value of each of the colors. Multiplying the chroma by the value of yellow 7/9, $7 \times 9 = 63$, and doing the same with purple-blue 3/4, $3 \times 4 = 12$, we get these two products 63 and 12. These are applied inversely, as in the former case, and we use 63 parts of purple-blue 3/4 with 12 parts of yellow 7/9. The conclusion is that the *stronger chroma and higher value should occupy the lesser area and the weaker chroma and lower value should occupy the greater area.*

COLOR COMBINATIONS

If in the foregoing we have touched upon the combining of colors in use, it has been only by way of explanation of some point in the laws of measurement and balance; and it is hoped that no impression has been created that the color combinations possible within the range of the Munsell System are limited to the examples which have thus far been mentioned. This is so far from being the case that any attempt to cover the subject of color combinations possible to this System would be quite futile within the limited scope of this chapter. A logical and orderly system will, in fact, offer a greater range of possibilities for the combination of color than could be discovered at random.

We must, therefore, be content to mention here only a few of the directions or paths which offer harmonious color combinations, trusting that the reader may be sufficiently interested by these to seek other possibilities of his own accord.

In considering the use of two colors together, we have repeatedly alluded to those having opposite hues, because this appeared to be the clearest example with which to explain the idea of balance. This combination of opposites is one of the simplest and surest of color harmonies. We have seen how, if properly proportioned as to amount or area, these opposite colors will balance in perfect neutrality; but another interesting fact with regard to them is that when placed together these contrasting colors tend to stimulate and enhance each other.

Another very simple and practically infallible series of color harmonies may be made within a single hue. (See Figure 19.) Thus we may combine a low value of any hue with a high value of the same; or, a weak chroma of any hue with a stronger chroma of the same. A more interesting combination within a single hue is that of a low value and weak chroma with a high value and stronger chroma or vice versa. Again, areas may be balanced, the stronger chroma having the smaller area and the weaker chroma the larger.

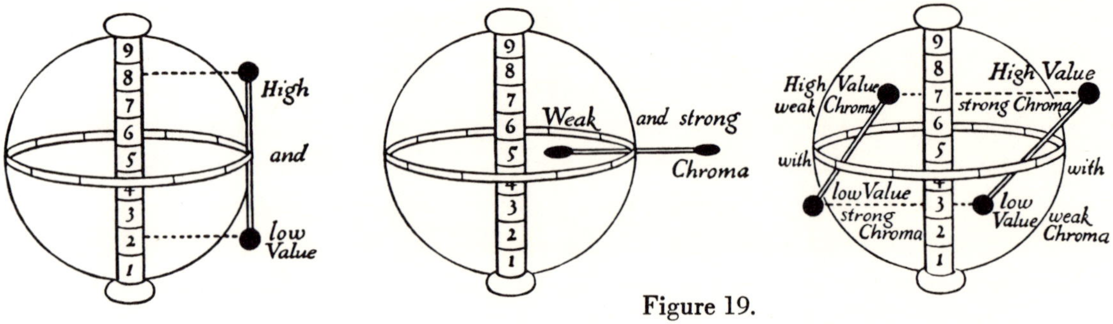

Figure 19.

Successful combinations can also be made between what are known as neighboring hues, that is of any hue with the hue which immediately precedes or follows it on the scale — green with green-yellow, red with yellow-red, yellow with yellow-red, etc. These may in turn be varied by taking them at different steps of value and different strengths of chroma. In the same way, hues may be combined with neighboring intermediary hues. In all of these cases the harmony depends upon proximity rather than contrast, as in the case of opposites.

The use of three or more colors will present a problem at once more complex and more interesting and which, if approached in any regular order, may assuredly be solved harmoniously. One method is to choose a certain restricted field of hues such as yellow to red, for example, and then to select within this field regular steps of hue, value, and chroma which bear an orderly relation to each other.

The principle governing the balance of opposite colors will also apply to combinations of three colors. Let us assume that blue is required as one of the colors in a three-color combination. We find that its opposite hue is yellow-red, and as this is merely an admixture of yellow and red it follows logically that the use of these two hues, with due regard to proportion of areas or strength of chroma, will yield a perfect color balance. (See Figure 20.) In order to de-

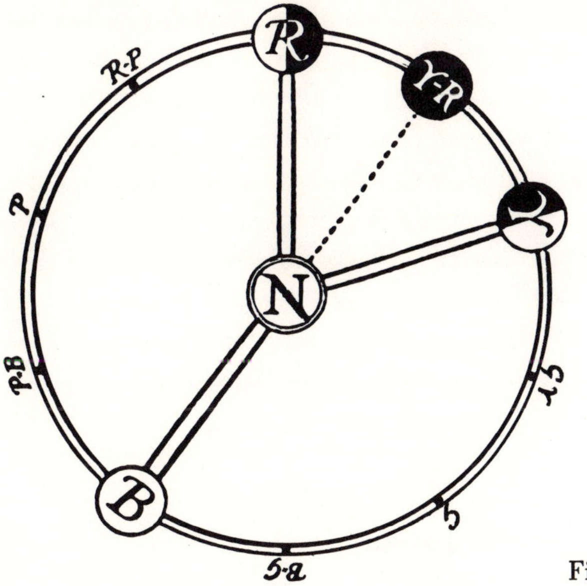

Figure 20.

termine the correct proportion of areas, or strength of chroma of red and yellow which will balance harmoniously with our blue, we may proceed exactly as in the case of a two-color combination of blue and yellow-red; but in this case we would divide the amount or strength of a correct yellow-red between our yellow and our red. For example, let us take blue 4/5 and assume that we wish to combine it with a yellow and a red of higher value and stronger chroma, say 6/7. Following the rule already stated, we multiply the value of our blue by its chroma, that is 4×5, which gives the product, 20. Now taking its opposite, yellow-red at 6/7, and doing the same we get $6 \times 7 = 42$. If we were combining blue 4/5 with yellow-red 6/7 we would use their products inversely, that is we would use 42 parts of blue 4/5 with 20 parts of yellow-red 6/7. This gives us the amount of area for yellow and for red, because if we would use 20 parts of yellow-red, 6/7, it naturally follows that we would use 10 parts of red 6/7 and 10 parts of yellow 6/7 to effect the same balance.

We may note one more interesting point which will be of value in connection with the use of several colors, two of which are of opposite hues. In studying the dimension chroma we have seen that all of the hues cross and meet in the neutral pole, which represents the point of their union. It follows naturally that the nearer our colors approach to this common center (the weaker they are in chroma) the more nearly they are related; and the easier it becomes to harmonize them. Now two of our hues being direct opposites will balance each other very well; but in the choice of other hues between these we shall be in danger of discord as we leave their immediate proximity and arrive at points

Figure 21.

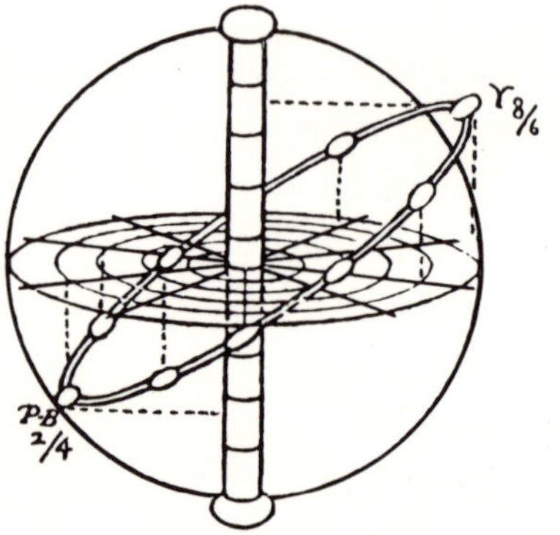

Elliptical path between opposite Hues of high and low Value

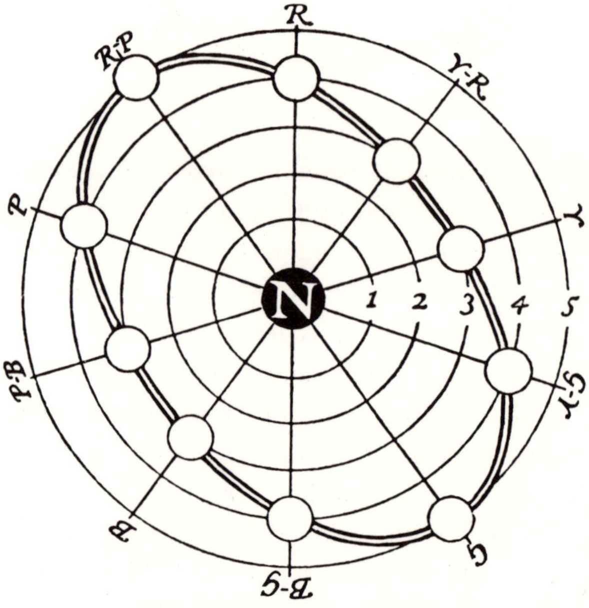

The elliptical path

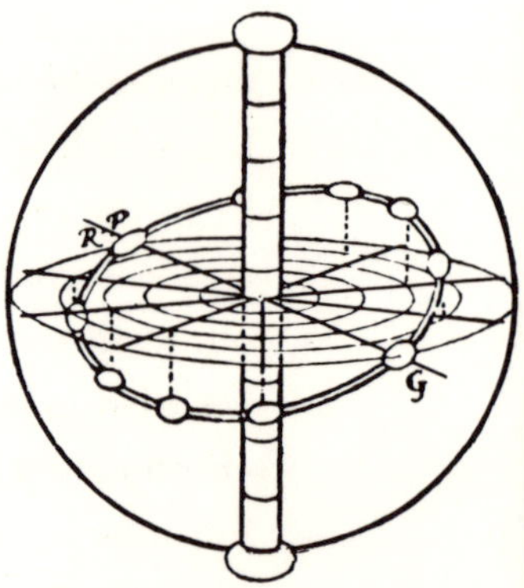

Elliptical path between opposite Hues of same Value and other Hues of high & low Value

half-way between them, where we find neither the balance of proximity nor of contrast. We may avoid this danger in the selection of our colors between these opposites by choosing steps of chroma for them which shall be nearer to the neutral pole and approach to within, let us say, three steps of it. The line thus traced between our opposite hues will form an ellipse and colors taken anywhere on this line will safely accord. This may be more readily comprehended by a glance at Figure 21. This suggests variations in the application of the rule, such as are indicated in the smaller perspectives, where the elliptical path is shown tilted to different levels of value.

A further study of color thus arranged in measurable order will assuredly be rewarded by the discovery of many interesting possibilities which we have failed to note here. The subject is endless and unless this chapter is to be likewise endless, the few suggestions which it offers must suffice. The deeper we penetrate this always fascinating subject, the more clearly we shall see that "color harmony" is only another term for color *order*; that order will yield order; and that any path in the Color Sphere, and some paths outside it, which are themselves orderly in form and interval, will lead through a series of colors which accord, and when used together will render the agreeable sensation which we seek in all color relations.

The Art of Harmony

Faber Birren has been responsible for the writing of the chapters that follow. While there may be some repetition of what has preceded, a well organized and fairly simple exposition of Munsell's ideas will now be attempted both as to color harmony and color notation. Also, various experiments will be proposed which the reader may wish to carry out. In effect, these final pages are meant to serve as a lucid and practical course of training for teachers and students, for amateurs and professionals in the fine and commercial arts, for designers, decorators, graphic artists.

As will be seen, Albert H. Munsell had a refined and sophisticated taste for color. Above all, he was insistent on neat order. Perhaps as a theorist and artist he stands opposite to that recent school of Abstract-Impressionism which sought color effects through impulse and individual feeling and which denied and rejected all formal training.

Today, however, there is a return to order, and the profound concepts and conclusions of Munsell take on new importance. For in Optical (Op) Art, in new color expression built around modern studies of human vision and perception, the Munsell System has a significant place. It restores to the field of art an old tradition in which knowledge and understanding are looked upon as essential background to creative effort.

Munsell was not much of a writer, and though he could think and speak clearly and coherently, to all indications he found the task of author a difficult one. The best presentations of his views have been by others. Although he wrote a number of papers and articles, his only book, *A Color Notation*, issued in 1905, when he was forty-seven, is academic and almost prosaic in its literary style. Munsell's achievements are far greater than he was able to put down in words. During the past few decades able scientists have come to his aid as to the technical perfection of his system of notation. And this book endeavors to do likewise in the realm of esthetics.

BEAUTY ACCORDING TO MUNSELL

Munsell's attitudes toward color harmony and beauty were never vague. He was a capable artist of conservative bent and he did not hesitate to differentiate the good from the bad as he saw them. If he was critical, however, he was also constructive, for he devised a most novel and impressive set of patterns and principles for the achievement of beauty. And he further invented ways by which such beauty could be anticipated, measured, and set forth in written formulas. If this sounds orthodox and rigid, and if it seems to run against the grain of the creative mind, Munsell is to be readily defended. Goethe once expressed the thought that rules were important to learn, if only for the privilege of breaking them. Munsell had a unique viewpoint, strict though it may have been. It is to the great benefit of any student of color, any designer or artist, to grasp the nature of these ideas, for, even if they are merely appreciated in a general way, they will help to point out directions of harmony and discord along which an artist may or may not choose to travel. At least, the creative impulse, the search for that which is singular and dramatic in color, will not be confused in a thick fog of ignorance.

Although Munsell as an artist painted in the heyday of Impressionism, he followed the conservative tradition and had too well-ordered a mind to be reckless or audacious with color. (See Figures 22 and 23.) He confessed his feelings again and again: "The circus wagon and poster, although they yell successfully for momentary attention, soon become so painful to the vision that we turn from them." In *A Color Notation* he wrote, "Beginners should avoid strong color." And again, "Quiet color is a mark of good taste." He didn't want color to be "loud." Refinement was imperative: "If we wish our children to become well-bred, is it logical to begin by encouraging barbarous tastes?"

Figure 22. THE HERRING NET, Winslow Homer (1836-1910). This American painter lived at the time of Albert H. Munsell and was also born in Boston. Both men respected the conservative tradition in art. (Photograph, courtesy The Art Institute of Chicago.)

Figure 23. ON THE TERRACE, Pierre Auguste Renoir (1841-1919). This French painter also was a contemporary of Munsell — and Munsell at one time studied in Paris. Yet Munsell showed little interest in the color theories of Impressionism. (Photograph, courtesy The Art Institute of Chicago.)

This was a stern attitude, but at the beginning of this century (1905), when Munsell painted and when he wrote *A Color Notation*, conservatism in art was the rule. Impressionism was beginning to win recognition, but radical schools of color in art, such as Fauvism and Orphism, were yet to come.

Abstract Expressionism, which burst upon the world of art during the middle of the century would probably have been disliked by Munsell. (See Figure

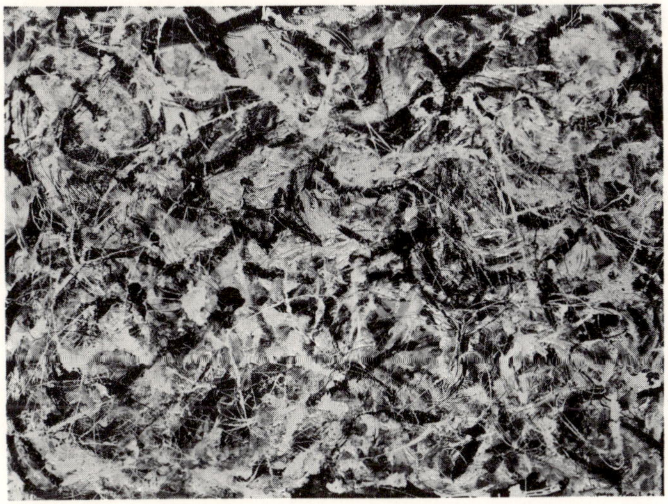

Figure 24. GRAYED RAINBOW, Jackson Pollock (1912-1956). Pollock was a prominent leader of the school of Abstract Expressionism and well represented the artistic credo of free expression — without discipline. (Photograph, courtesy of The Art Institute of Chicago.)

Figure 25. GROUND HOG DAY, Andrew Wyeth. Here are conservatism, tradition, and discipline in American art and color expression. Wyeth painted often in the New England of Munsell. (Photograph, courtesy The Philadelphia Museum of Art.)

24.) In Impressionism, in Fauvism, and Orphism, there was a certain measure of discipline and respect for knowledge and training. But in Abstract Expressionism the artist was to reject and renounce all that was formal. He was to give release to inner moods and impulses entirely. In some forms of action painting he was literally to fling his colors at the canvas before him, pour them on the floor, or wrap his feet with rags, soak them in paint and thus stomp away at an uninhibited composition.

THE NEED FOR BALANCE

Discipline is back these days, and Munsell's theories well deserve reappraisal. Abandonment to impulse in the use of color has passed over the horizon. The student of color today, the artist and designer, is demanding control over his medium. Beauty will be planned. Yet originality and progress will go on because of new references and sources of inspiration — the probing of phenomena associated with human vision and perception. (See Figure 25.)

To Munsell the most important of all factors in color harmony was BALANCE. This need be repeated again and again: "The use of strongest colors only fatigues the eyes, which is also true of the weakest colors. In a *broad* way one may say that color balances on middle gray." He was not entirely opposed to discord and unbalance. He believed that the wise use of discord could enhance the overall balance of a color harmony; it was the correction of discord by overall balance that led to a true art of color harmony.

In *A Color Notation* he pointed out that many color schemes in paintings could be criticized as being too light or too dark (unbalanced in value), too weak or too strong (unbalanced in chroma), or too hot or too cold (unbalanced in hue). What he meant by balance will shortly be explained. There must be reason and consistency: "A measured and orderly relation of color underlies the idea of harmony." An ingenious or sensitive mortal might be born with a fair sense of color, or he might by sheer psychic endowment create remarkable effects, but the surest road was one of struggle and comprehension: "Appreciation of beautiful color grows by exercise and discrimination, just as naturally as fine perception of music or literature. Each is an outlet for the expression of taste — a language which may be used clumsily or with skill."

As will soon be emphasized, Munsell liked middle value colors and particularly middle gray N5. (See Figure 26.) Middle value colors were those which in lightness and darkness appeared to the eye to lie midway between black and white. Still further (as will be seen), he favored middle value colors

Figure 26.

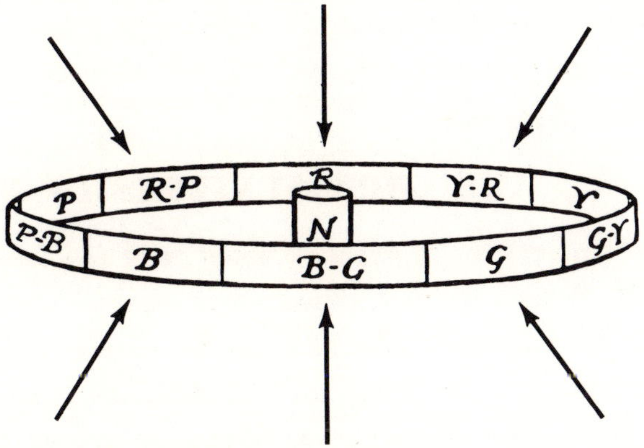

having a soft or muted purity (chroma). Thus 5 for neutral gray, or 5/5 for hues, became special notations for Munsell. It was the 5/5 colors (of medium 5/ value and /5 degrees of chroma) that were produced in crayons, water-colors, oil paints, papers, and which became something of a vogue in color education during the twenties and thirties. (Also included in such art materials were the same colors or hues at maximum chroma.) Munsell liked 5/5 combinations of opposites, for equal proportions of them spun and visually "mixed" on an electric motor would form neutral gray N5 — and this was the height of refinement and elegance.

In a similar way, light colors dominated dark colors, and where they were used together it was best to have a sequence between them also pass through N5, or at least through middle value.

Remember that in Munsell's day, Victorian survivals of conservatism still existed. In painting, design, home furnishings, interior decoration, most people demanded restrained taste. Balance was perhaps sought because it led to that which was subtle and exquisite. However, a color revolt came which suddenly and in a spectacular way led to extreme vividness of color. While such blatant effects were allowable and could be plotted within the confines of Munsell's theories, they were by no means looked upon with much tolerance.

Now there are indications of a reaction and return in favor of balance, and what Munsell invented in the esthetics of color control may well gain renewed attention and prestige.

THE CHIEF FEATURES OF BALANCE

Munsell's principles of color harmony will now be discussed and illustrated in detail. Here is a brief introductory summary of them. They are nine in number. He was quite liberal as to the choice of key *hues*, even though he favored opposites. He permitted and recognized the use of adjacent or neighboring hues, split-complements, triads, or indeed all the hues of his color circle, but carefully organized.

(A) Gray colors harmonize best when neatly and evenly spaced with reference to the 9 steps of the gray scale. Harmonies based on N5 as a key or center point are preferred. (See Figure 27.)

(B) Monochromatic harmonies (using one hue only with variations achieved by mixing them with black and white) also look best when there is neat spacing as to value *and* chroma. A key or center point of balance is preferred at middle value 5. An example would be R 7/5 with R 3/5.

(C) Opposite colors of medium chroma /5, and which find sequences in middle gray N5, are harmonious when combined in equal area.

(D) Opposite colors of equal value (preferably value 5/) but of different degrees of chroma are also beautiful if the color of weaker chroma is given larger area than the color of stronger chroma. An example would be R 5/10 with a twice larger area of BG 5/5.

(E) Opposite colors of the same chroma but of different value will harmonize if neatly spaced with reference to the gray scale. They would look best if the point of balance was N5. An example would be R 7/5 with BG 3/5 or R 8/4 with BG 2/4.

(F) Opposite colors of different value and different chroma will harmonize if neatly spaced with reference to the gray scale and if the deeper color or

Figure 27.

A

N1, N3, N5, N7, N9

B

N5, N7, N8, N9

C

N1, N3, N4, N7

D

N3, N5, N7, N9

E

N1, N3, N5, N7

weaker chroma is given larger area than the color of lighter value or stronger chroma. An example would be R 7/8 with BG 3/4.

(G) Colors not complementary to each other, such as neighboring hues or split-complements, will harmonize if the above principles are followed. Such colors held to the same value and chroma will automatically please the eye. Examples would be YR 5/5 with R 5/5 and Y 5/5 for neighboring hues, or R 6/10 with B 4/5 and G 4/5 for split-complements. Or one color may have high value and one low value, with the third color held at middle value. An example for neighboring hues would be YR 7/7 with R 5/5 and RP 3/3. Here the lighter value should have smaller area than the middle value and deep value. An example for split-complements would be GY 7/10 with RP 5/8 and PB 3/6.

(H) Where there were ranges of color, Munsell spoke favorably of "diminishing sequences." As an example, starting out with a light orange Y 8/9, a diminishing series of steps was created which dropped one value step and one chroma step as it proceeded. Thus Y 8/9 went on to GY 7/8, G 6/7, BG 5/6, B 4/5 to PB 3/4. As he wrote, "Nature seems to be partial to this sequence."

(I) Finally there were elliptical paths to be drawn. (See Figure 21 and the Munsell System of Color Combinations described by T. M. Cleland in the previous chapter.) As one color was scaled to its complement, the intermediate steps could trace paths that moved in neat steps toward and away from neutral gray, either on the same level of value or between different levels. This principle will be dealt with more fully later.

The above Principles will now be described in greater detail and many of them will be illustrated in black-and-white and full color.

Perhaps it should be emphasized at this point that while Munsell's ideas of balance were unique with him and an important contribution to the art of color harmony, he by no means insisted on balance alone to the exclusion of other color arrangements — or even discord and unbalance. A footnote in the eleventh edition of his *A Color Notation* observes: "A careful study of Mr. Munsell's writings proves beyond the slightest trace of doubt that his ideas on 'Balance' were far broader than often has been assumed. Mr. Munsell considered 'Balance' about neutral gray, merely as the simplest and most elementary form of balance, which should be mastered by the thorough student of color before going on to the more interesting centers of balance." Mrs. B. R. Bellamy, manager of the Munsell Color Company in Baltimore, further remarks, "Mr. Munsell laid down certain specific rules to follow to assure balance but believed after they were learned thoroughly, they need not be followed rigidly but altered to suit the artist and the subject he wished to portray. He also pointed out the monotonous use of a single value, or a constant value/chroma combination."

In the principles, discussions, and illustrations that follow, Munsell's more orthodox principles of balance will be duly described and shown, but added to them will be color arrangements which break with the academic and, in broader scope and range, suggest freer, more imaginative flights of fancy.

PRINCIPLE A, THE BALANCE OF NEUTRAL GRAYS

Refer to Figure 27. Neutral gray N5 is the central point of balance for the entire Munsell System. Above and below it are higher and lower values. To the side are stronger chromas. Munsell felt that N5 was something of a basic headquarters or depot from which color ranges followed tracks to pure hues, to white and to black. Some modern psychologists look upon gray as a primary sensation. Certainly there are few, if any, persons who, upon looking at a gray area, will describe it as being a whitish black or a blackish white.

The gray scale itself was harmonious. And so were any neatly related steps. For example, the combination of N3, N5, N7 was harmonious with black on a white ground, as in A of Figure 27. The eye sensed that all was in order and in proper balance.

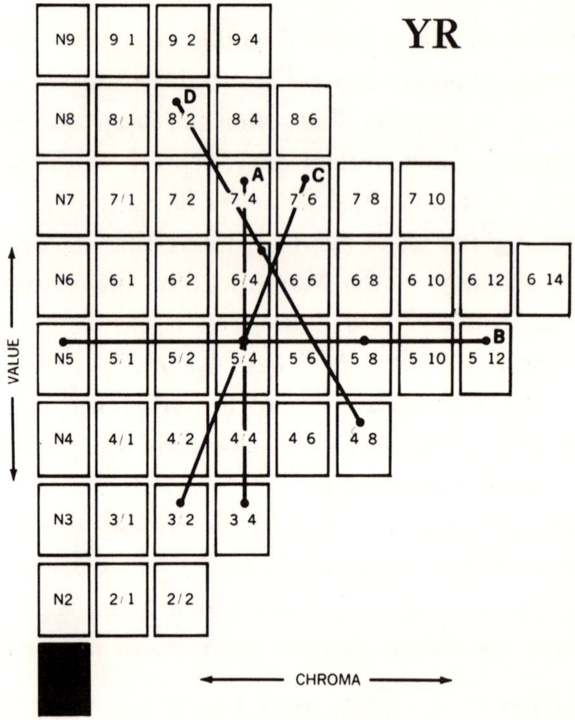

Figure 28.

However, if one were to combine N5 with N7 and N8 on white, or N3 and N4 with N7 on black, as in B and C of Figure 27, something would be amiss.

It was best to have one value relate to another in direct rather than haphazard order, as in D and E of Figure 27. Depending on a particular design or composition, the matter of area for different values was not too significant. Normally, light values are more stimulating than dark values. So if dramatic emphasis were wanted — with good balance — the light values could be restricted in size and the dark values given broader distribution.

Yet colors also have apparent weight, black being "heavy" and white "atmospheric." Hence in a generally light composition, small accents of black or a deep value could be made to appear dynamic, and also in balance.

PRINCIPLE B, MONOCHROMATIC HARMONIES

Every vertical slice through the Munsell Color Solid is composed of values and chromas of a single hue. Color Plate II shows such a chart for yellow. Figures 28 indicates, in black and white, the normal hues and values for yellow-red (orange), while Plate III shows color harmonies as described herewith. It will be noted in Figure 28 that the first chroma step is /1. This is followed by even steps of /2, /4, /6, /8, /10, /12, /14. Value steps run consecutively from 2/ to 9/.

First working up and down in a value sequence, if YR 5/4 may be taken as a more or less central point on Figure 28, this medium soft chroma harmonizes effectively with YR 3/4 and YR 7/4 on a white ground and with black outlines. (A on Color Plate III.)

In a horizontal chroma sequence (B on Color Plate III) and using Munsell's favorite middle value, a bright yellow-red (orange) 5/12 looks well with 5/8 and 5/4 on a neutral N5 ground, even though the tones melt together. (It would be permissible here to add black and/or white inasmuch as these two extreme values are in balance with N5.)

In diagonal sequences, orange colors (YR) of high value and strong chroma can be combined with orange colors of dark value and weak chroma — but preferably in sequence with a middle value tone. Or a color of high value and weak chroma can be combined with a color of dark value and strong chroma. The former effect will be seen in C on Color Plate III. Here 5/4 is again taken as a mid point and combined with 7/6 and 3/2. The light tone is relatively pure, while the deep tone is more neutral and toward black.

In D on Color Plate III, another diagonal sequence is shown, this one from 8/2 to 6/4 to 4/8. (Also see Figure 28 on which the above sequences are plotted. This chart is based on the 1950 edition of *The Munsell Book of Color*.)

Exercises like this can, of course, be carried out with any hue. Equally spaced steps in vertical, horizontal, or diagonal lines are delightful to behold — especially, Munsell held, if one of the steps is firmly planted on a middle 5 value.

(Note: Color sections or charts in *The Munsell Book of Color* will be found to vary in different editions. This will present no great problem in working out the color effects described in these pages and included on the Color Plates. For purposes of color harmony it will be perfectly satisfactory to use the eye and good judgment in establishing value levels and chroma steps which do not happen to be found in color chips on actual Munsell charts. Further, while Munsell during his lifetime made a big feature of value 5, chroma 5 colors in discussing beauty with color, these particular tones are *not* to be found in recent Munsell charts which give chroma sequences as 1, 2, 4, 6, 8, 10, 12, 14, etc. Where 5/5 tones are called for, these are very easily determined by forming colors that *appear* to be between 5/4 and 5/6. As a matter of historical record, in an *Atlas* originally designed by Munsell to illustrate his System, chroma steps from 1 to 10 *inclusive* were for the most part exhibited. However, as the System was improved and extended after 1929 — by others — only even chroma steps, 2, 4, 6, 8, etc., were shown in actual sample, this being the form in which *The Munsell Book of Color* appears today.)

PRINCIPLE C, OPPOSITE COLORS OF EQUAL CHROMA THAT CENTER ON MIDDLE VALUE N5

This principle was favored by Munsell and was one which he deservedly considered to be original with him. It gave him a chance to work with light and dark colors, to feature his middle value, medium chroma colors — and to prove beauty in mathematical terms.

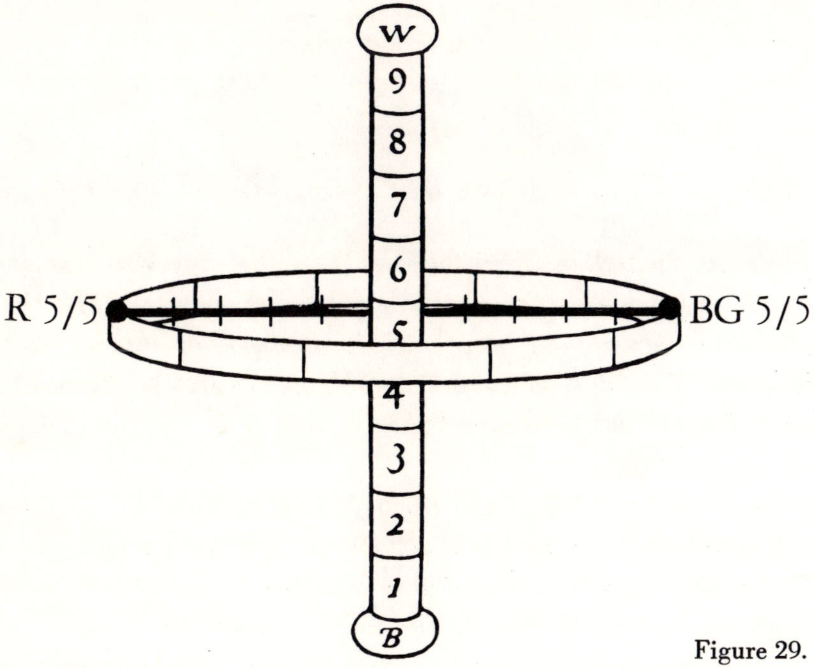

Figure 29.

There are five pairs of opposites on the Munsell ten-color Hue Circuit: red and blue-green, yellow and purple-blue, green and red-purple, blue and yellow-red (orange), and purple and green-yellow. Each has a different quality. Red and blue-green combine great warmth with coolness. Yellow and purple-blue combine a luminous color with a subdued one. Green and red-purple seem to balance warmth and coolness. Blue and yellow-red is visually dynamic. Purple with green-yellow seems exotic.

Illustration A on Color Plate IV shows red 5/5 with blue-green 5/5 on a neutral gray ground of middle value N5. (Also see Figure 29.) This may appear unusually conservative and suppressed in a modern world of bright color, but to Munsell and in Munsell's day, a scheme like this exemplified

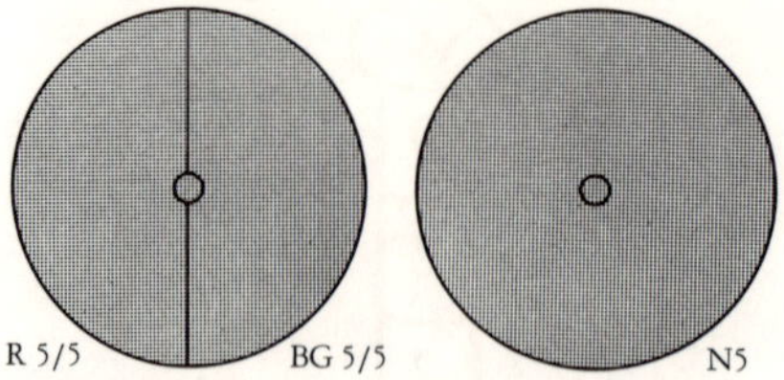

Figure 30.

true elegance. Munsell realized this and therefore suggested that one of the hues could be a step higher than middle value and one a step lower than middle value, *but they should connect in "a sequence of chroma balanced on middle gray*." Thus in B on Color Plate IV, the N5 gray background remains the same. The red has been *raised* from 5/5 to 6/5, and the blue-green has been lowered to 4/5.

If Maxwell disks of R 5/5 and BG 5/5, or R 6/5 and BG 4/5, were combined in equal 50-50 per cent proportions and spun on a motor, the resultant visual mixtures would be a neutral gray N5. (See Figure 30.) Here Munsell's ideal balance would be achieved. The formula would be an esthetic one and would verify his contention that "a measured and orderly relation underlies the idea of harmony."

As exercises for Principle C, it seems best to have one color higher in value than the other, but the balance must be on middle gray 5N and the chromas must be medium. As plotted on Figure 31 highly pleasing schemes are to be found with the following combinations (5N gray can be added or not, as desired): Y 6/5 with PB 4/5, or Y 7/5 with PB 3/5. Similar harmonies are to be found in RP 6/5 with G 4/5, or RP 7/5 with G 3/5. Or try YR 6/5 with B 4/5, or YR 7/5 with B 3/5. Or try GY 6/5 with P 4/5, or GY 7/5 with P 3/5.

All the above combinations would be on the muted side, for all have a medium /5 chroma. All could be charted as on Figure 31. (Brighter effects will be discussed in principles that follow.)

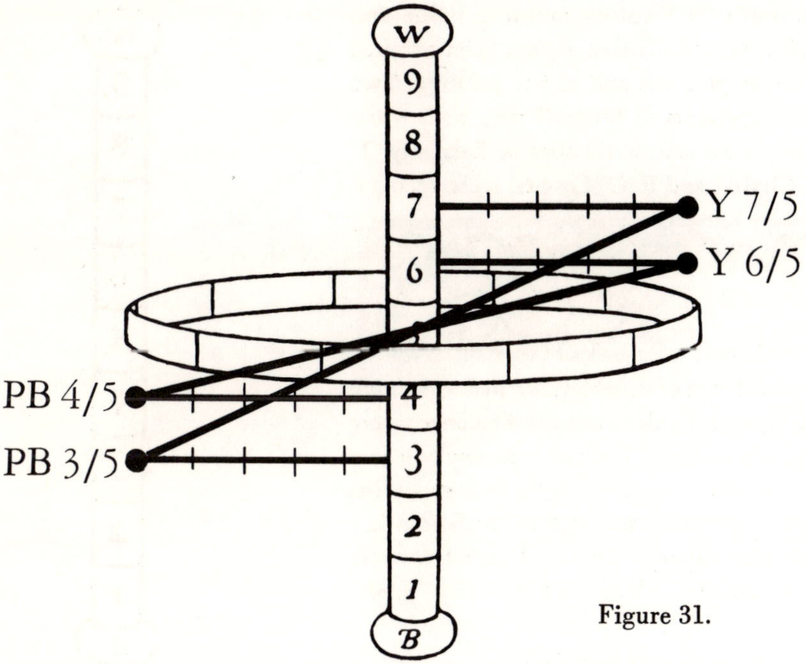

Figure 31.

Munsell's 5/5 colors (middle value and medium chroma) have always been admired in the graphic arts and in fine printing. Two loyal exponents of Munsell who both wrote monographs on his theories of harmony, T. M. Cleland and F. G. Cooper, were typographers and type designers. Color in fine printing and typography demands good balance where color is introduced. For example, any type face has apparent "weight," light-face types (as used in this book) are "lighter" than bold-face types which appear "heavy." Thus white paper is "weightless," while black is totally "solid."

Typographers have a rule that bold-face types in color balance best with light-face types in black. And middle-value colors, those that lie apparently halfway between black and white, look beautiful when combined with black type on white paper.

Munsell's 5/5 colors (middle value and medium chroma) have always been admired in the graphic arts and in fine printing. Two loyal exponents of Munsell who both wrote monographs on his theories of harmony, T. M. Cleland and F. G. Cooper, were typogra-

phers and type designers. Color in fine printing and typography demands good balance where color is introduced. For example, any type face has apparent "weight," light face types (as used in this book) are "lighter" than bold face types which appear "heavy." Thus white paper is "weightless," while black is totally "solid."

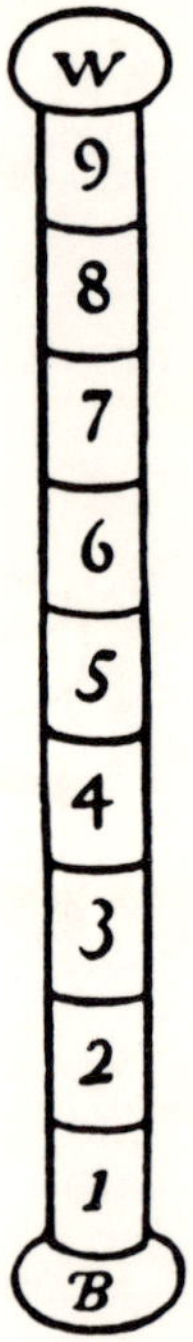

Figure 32.

In C on Color Plate IV (and in Figure 32) this quality of color balance is illustrated. Middle value colors, lying halfway between black and white, do not have to be of medium chroma. A brilliant red and purple-red will be found at value 5/, and so will a fairly pure green, blue-green, and blue. However, Munsell's 5/5 colors themselves suggest refinement and come highly recommended to printers and designers in the graphic arts.

PRINCIPLE D, OPPOSITE COLORS OF EQUAL VALUE BUT OF DIFFERENT CHROMA

In this principle Munsell preferred to work at middle value N5 for absolute balance. However, he was willing to admit the use of opposite colors of differ-

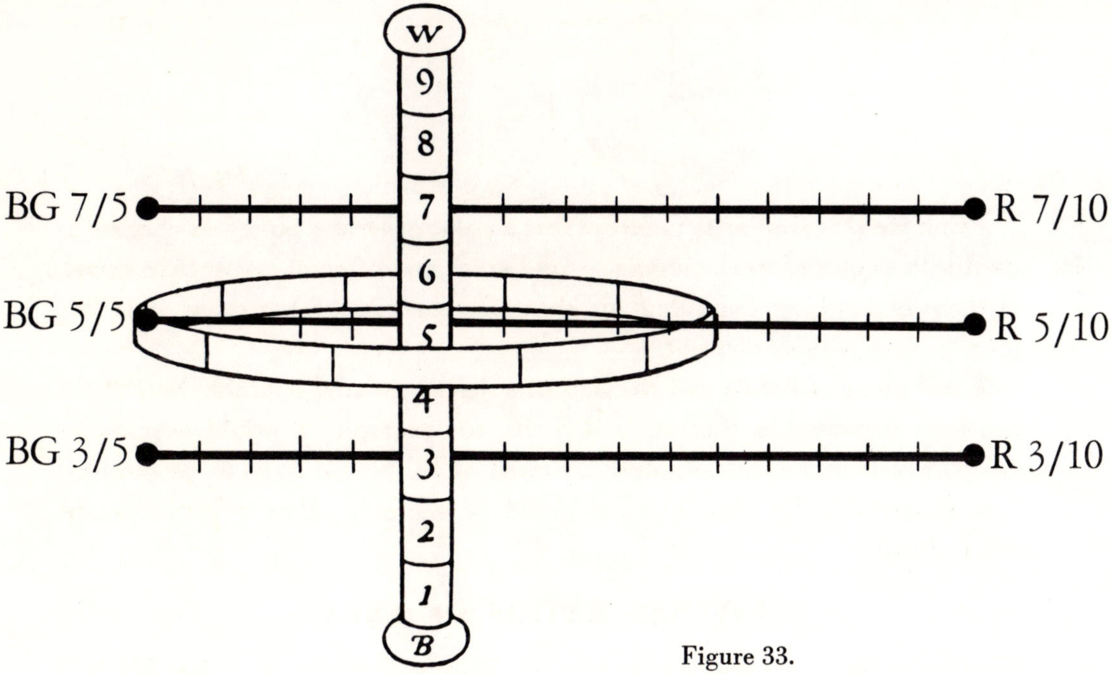

Figure 33.

ent degrees of chroma at other values from N9 down to N2. In all cases, however, the opposite colors in proper area were expected to cancel into neutral gray if such areas were measured and spun on an electric motor — whether such gray was light, medium, or deep. (See Figure 33 which plots three combinations of red and blue-green.)

It should be recognized, however, that combinations of colors having the same value may appear extremely conservative, if not monotonous. Thus Principle D is a good one to follow in order to gain knowledge and practice in matters of color balance. Yet better principles having more variety and excitement, and with wider application, will be found in E, F, G, H, and I.

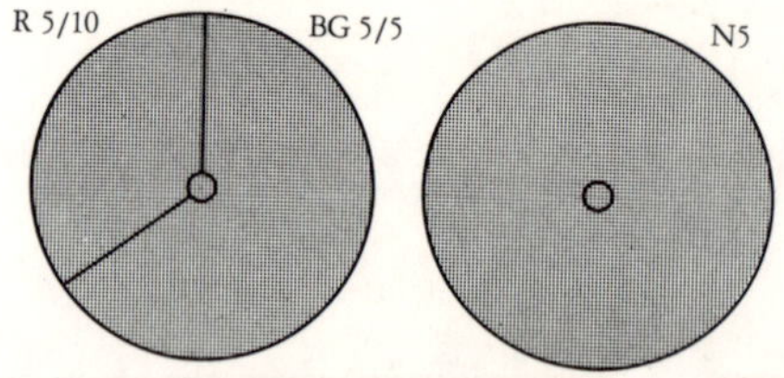

Figure 34.

Principle D is similar to C except that a color of strong purity or chroma is used with a color of weak chroma — and areas are different rather than equal. If there is anything individual to Munsell's theories of harmony, it is this more or less positive concept.

Equal areas of R 5/5 and BG 5/5 in a design assured balance. Now if the red were increased in chroma to R 5/10, for example, it would overwhelm BG 5/5. To correct this situation the area of BG would have to be doubled (see Figure 34) for then balance would be achieved. How might areas be determined?

INVERSE RATIOS OF AREA

Munsell spoke of inverse ratios of area. To return to the R 5/10, BG 5/5 scheme, 5 × 10 (value and chroma) for the red (R) would equal 50, while 5 × 5 for the blue-green (BG) would equal 25. These proportions 50 and 25 were to be reversed for the two colors, with 50 being assigned to the blue-green and 25 to the red, or 2 to 1. In effect (as in Figure 34) 2 parts of blue-green 5/5 (66⅔%) and 1 part (33⅓%) red 5/10 if spun with Maxwell disks on a motor would form neutral gray N5! Beauty and concord were thus proved.

Munsell's inverse ratios of area were given wide application by him and will be brought up again in following discussions devoted to Principles E and F. To return to Principle D, which deals with opposite colors of equal value but of different chroma, here are a few suggested exercises. Balanced schemes at any level of value (but with middle value 5/ preferred) can be arranged. Assume a combination of yellow-red (YR) 7/6 with blue (B) 7/4. To figure

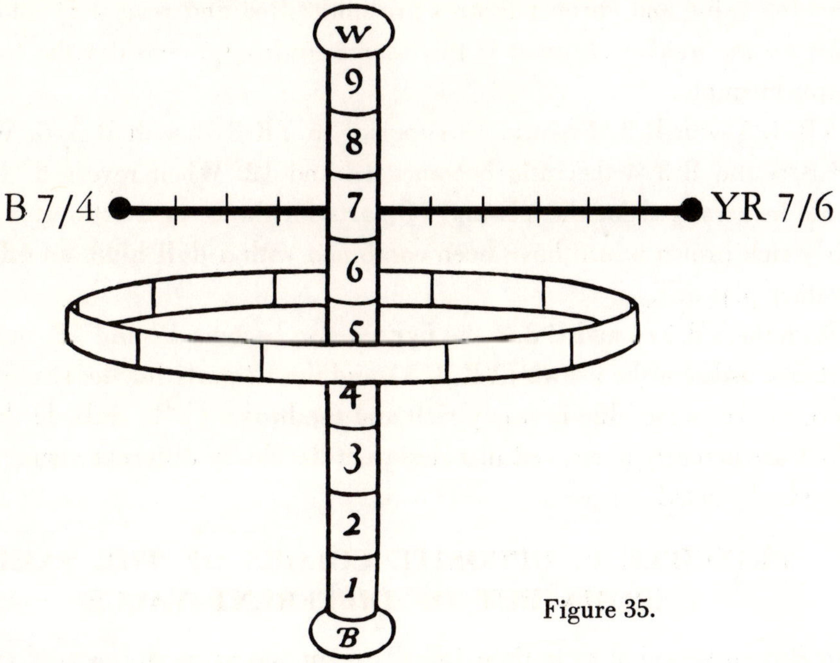

Figure 35.

balancing areas, the 7 × 6 would equal 42, and the 7 × 4 would equal 28. (See Figure 35.) Reversing the figures, 28 parts of YR 7/6 would balance 42 parts of B 7/4, a ratio of 3 to 2, and this ratio if measured in Maxwell disks on a motor would form a light neutral gray at N7.

The same procedure would hold true if the two colors were set at a lower value, such as 3. Thus yellow-red (YR) 3/6 with blue (B) 3/4, would produce figures of 18 and 12, or 3 to 2. With areas in reverse, 3 parts blue 3/4 would balance 2 parts YR 3/6 in a design (or if spun on a color wheel).

Principle D is a fairly easy one to comprehend. Opposite colors must be employed, and they must be of the same value. Only chromas are to differ. *As a matter of experience and observation, it is usually best to use a warm color for the stronger chroma* (which will automatically take the smaller area when the value and chroma figures are multiplied and reversed) *and a cool color for the weaker chroma.* If this seems confusing, consider the following simple example.

YR 3/6 with B 3/4 would be superior to YR 3/4 with B 3/6. With the YR 3/6 and B 3/4 the ratio becomes 18 and 12. When reversed, 12 parts would be arranged for the YR and 18 parts for the B, or a ratio of 3 to 2. A fairly rich brown would have been combined with a dull blue, an effect that is rather pleasing.

With the YR 3/4 and B 3/6 the figures also become 12 and 18, but the 18 parts are used for the brown (YR 3/4) and the 12 parts for the stronger blue (B 3/6). Here the blue is fairly rich and the brown (YR) dull. If these two effects are actually portrayed in a design, a decidedly different visual impression will be noted.

PRINCIPLE E, OPPOSITE COLORS OF THE SAME CHROMA BUT OF DIFFERENT VALUE

This Principle is similar to Principle C except that value differences are great rather than uniform or small. In Principle C opposite colors of the same chroma were contrasted at middle gray N5 or located slightly above and below at values 4/ and 6/. In Principle E, chromas are still held uniform but one color is made light and one dark — with a balance point again held at N5.

For example, red (R)7/5 will combine harmoniously with blue-green (BG) 3/5. Or R 8/4 can be combined with BG 2/4. These value differences, it will be noted, are equidistant from N5. And equal areas on a color wheel would cancel into gray.

As exercises, try YR 7/6 with B 3/6, or GY 7/6 with P 3/6. (See Figure 36.) Or try Y 7/12 with PB 3/12, and so on around the color circle.

Some interpreters of Munsell have recommended inverse ratios of area for colors of different value, suggesting that the lighter value be given the smaller

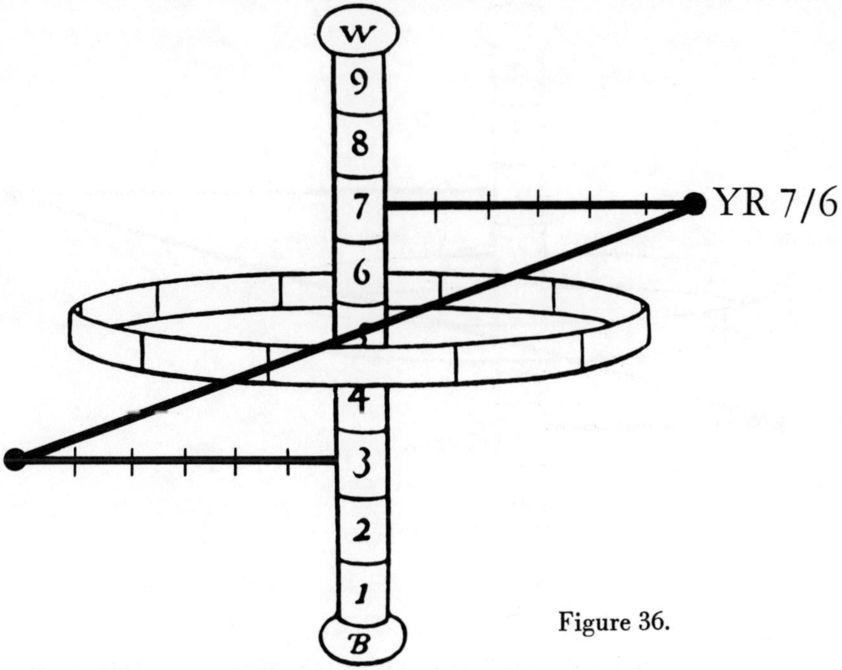

Figure 36.

area and the darker value the larger area. For example, in a combination of R 7/5 with BG 3/5 the 7 × 5 equals 35 and the 3 × 5 equals 15, a ratio of 7 to 3. Reversing matters, BG (blue-green) would be given an area of 7 parts to 3 parts for R (red). While this will appear harmonious, the inverse ratio idea does not need to be followed. A vast number of color schemes and compositions in the fine and commercial arts are executed on white paper. Such white, with a value of N9 can be made part of Principle E. As an instance, BG 3/5 with R 7/5 with white N9 will offer three neat value steps — and the areas can be held largest for white N9, smaller for R 7/5 and still smaller for BG 3/5.

However, Munsell's insistence that, with colors of equal value, the stronger chroma should have smaller area than the weaker chroma is remarkably true and esthetically sound.

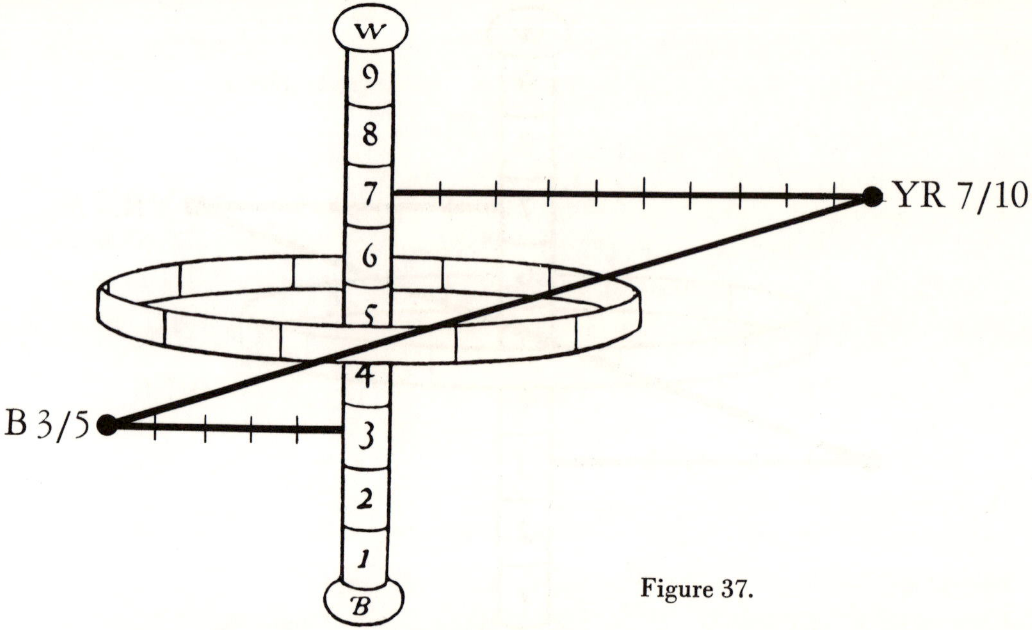

Figure 37.

PRINCIPLE F, OPPOSITE COLORS OF DIFFERENT VALUE AND DIFFERENT CHROMA

This Principle offers an opportunity to work with vivid colors, but still to preserve balance and to plan sequences that pass through middle gray N5. In all combinations it is best to have the values be in regular steps above and below N5 and the chromas to be similarly arranged in neat steps from the gray scale. Here are examples:

Yellow (Y)8/12 with purple-blue (PB)2/6. Or yellow-red (YR)7/10 with blue (B)3/5. (See Figure 37.) Or green-yellow (GY)7/10 with purple (P) 3/5. Or red-purple (RP)6/8 with green (G)4/4. All these schemes are shown on Color Plate V which should be consulted. All combinations find sequence through middle gray N5. Again it is best (a) to have the warmer color be stronger in chroma and lighter in value, with (b) the cooler color weaker in chroma and lower in value.

Inverse ratios of area work well with Principle F. In the instance of YR 7/10 with B 3/5 plotted in Figure 37, 7 × 10 equals 70, and 3 × 5 equals 15, or a ratio of 14 to 3. Thus the area for YR (reversed) would be 3 parts and the area for B would be 14 parts. Mathematical formulas like this are perhaps academic and do not have to be literally carried out. The chief lesson to learn is that *in general* vivid colors ought to have smaller area than grayish colors when they are used together.

PRINCIPLE G, THE HARMONY OF NEIGHBORING HUES AND SPLIT-COMPLEMENTS

A number of studies in the field of esthetics and the psychology of beauty have revealed the fact that the color combinations most liked by average persons comprise either (a) colors closely related or (b) colors in marked contrast. Munsell agreed with this. His first preference was for exact opposites. He then spoke in favor of neighboring hues, split-complements, and lastly of broad ranges of hue which more or less embraced the entire color circle.

Neighboring colors are those that lie next to each other on the color circle: red with yellow-red and red-purple; yellow with yellow-red and green-yellow; green with green-yellow and blue-green; blue with blue-green and purple-blue; purple with purple-blue and red-purple, etc.

While these neighboring colors could be arranged in innumerable ways, Munsell recommended the following:

— Neighboring colors should be neatly stepped as to value and should find sequence at middle value 5.

— They should be of the same chroma for good balance.

— Or one color, preferably warm, could have strong chroma and the other color, preferably cool, could have weak chroma. Where chromas differed, the stronger should have less area than the weaker, and this could be determined by applying inverse ratios of areas previously described.

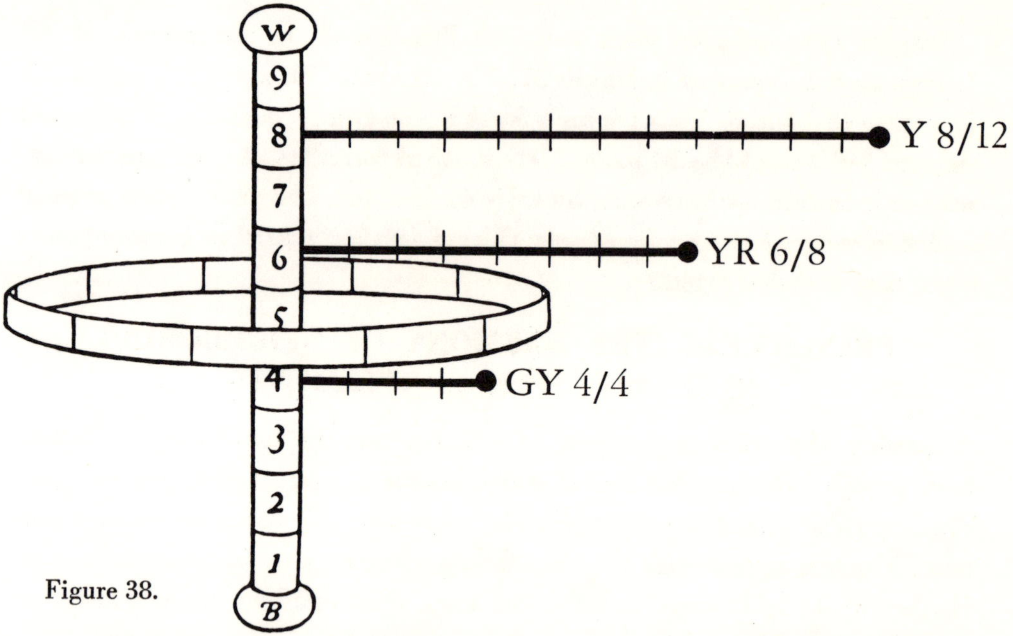

Figure 38.

Here are some examples. Purple (P) 5/5 combines well with red-purple (RP) 7/5 and purple-blue (PB) 3/5 — and all have equal chroma centering on middle gray. This is shown in illustration A on Color Plate VI.

Yellow (Y) 8/12 combines well with yellow-red (YR) 6/8 and green-yellow (GY) 4/4. This is a series of diminishing chromas and is shown in illustration B of Color Plate VI. (Also see Figure 38.)

The designer and artist can work out similar schemes throughout the color solid, and always with full assurance of success if the instructions given immediately above are followed.

With split-complements, a key color is combined with the two hues that lie next to its direct opposite. Red and yellow are the split-complements of blue, for example. Red-purple and yellow-red are the split-complements of blue-green. Yellow and green are the split-complements of purple, and so forth.

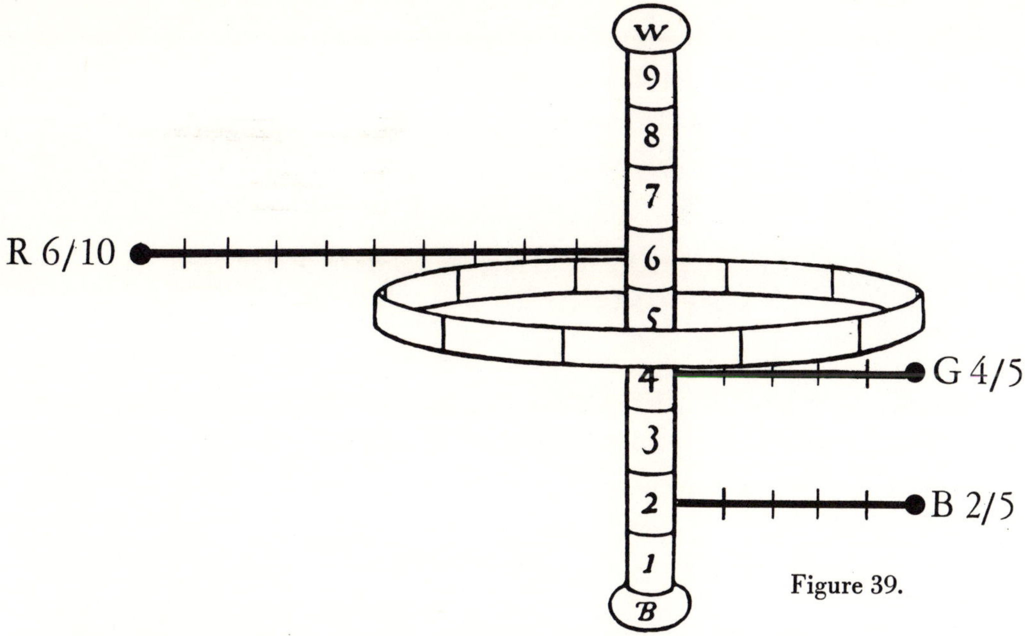

Figure 39.

As with neighboring colors the values and chromas should be neatly stepped. Warm colors make good key hues, with cool colors used for the split-complements and for background elements in a composition. Also the warm colors are best given strong chroma and the cool colors weak chroma. Inverse ratio of area can be determined as well.

In illustration C on Color Plate VI red is shown with its split complements blue and green. (Also see Figure 39.) Red is set at 6/10 and the blue and green are set at 2/5 and 4/5. If ratios were to be figured, the 6/10 for red would equal 60, and the 4/5 for green and 2/5 for blue would equal 30, a ratio of 2 to 1. Putting this in reverse 2 parts green and 1 part blue (a total of 3 parts) would balance $1\frac{1}{2}$ parts red. This combination, if spun on a color wheel, would form middle gray N5.

Another combination of split-complements is shown in illustration D on

Color Plate VI. This is composed of green-yellow (GY) 7/10, with red-purple (RP) 5/8, and purple-blue (PB) 3/6.

As with each of Munsell's nine Principles, an almost endless array of striking effects is to be achieved, and all in ideal visual balance if Munsell's excellent advice is followed.

PRINCIPLE H, THE HARMONY OF DIMINISHING SEQUENCES

Munsell stated that paths on the Color Sphere could trace laterally for hue, vertically for value, and horizontally for chroma. And all three movements could be made at one and the same time in what he called diminishing sequences. In describing the color of a buttercup and its complement, he noted a diminishing series of hues beginning with a bright yellow (Y) 8/9 and proceeding to green-yellow (GY) 7/8, green (G) 6/7, blue-green (BG) 5/6, blue (B) 4/5, to PB 3/4. The scale, as it proceeded from yellow to purple-blue, decreased evenly by one value step and one chroma step as it went from light to dark.

This color scheme is shown in illustration A on Color Plate VII and is diagrammed in Figure 40. It makes a circuit of the cool side of the sphere, starting with yellow (Y) and ending with purple-blue (PB).

A circuit of the warm side of the sphere is shown in illustration B of Color Plate VII, *and this also starts with yellow* (Y) 8/9 *and ends with purple-blue* (PB) 3/4.

Sequences like this are common in nature, as Munsell observed. As exercises, the artist or designer may start out with a key color of virtually any hue, of any value and chroma. From this point he can then develop a series of related tones (sequences) by shifting in neat steps toward a neighboring hue, toward a higher or lower value, and at the same time toward a stronger or weaker chroma, bearing out the "law" that order and balance equal harmony.

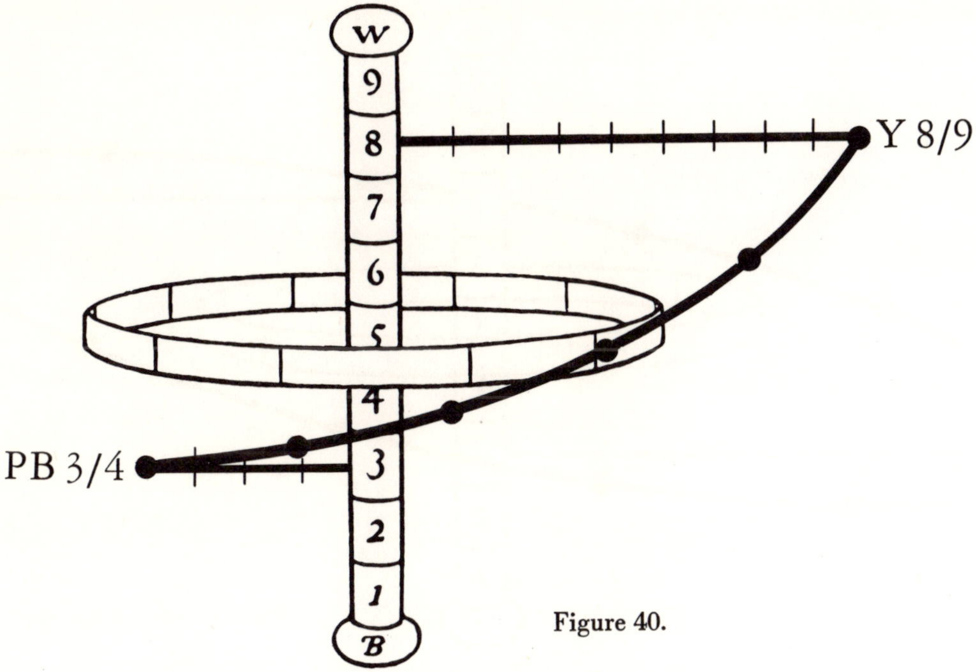

Figure 40.

PRINCIPLE I, THE HARMONY OF THE ELLIPTICAL PATH

As an advanced principle, Munsell spoke of triple balance and of elliptical paths which brought together various hues, values, and chromas all related in orderly sequences and with diminishing (or increasing) steps as described above with Principle H.

Elliptical paths could be arranged in one of three ways:

First, with opposite hues of the same value and with intermediate hues also on the same value but diminishing (and increasing) in chroma toward neutral gray. Here is an example charted by T. M. Cleland in the central diagram of Figure 21. All colors and modifications are at middle value 5. Two opposites,

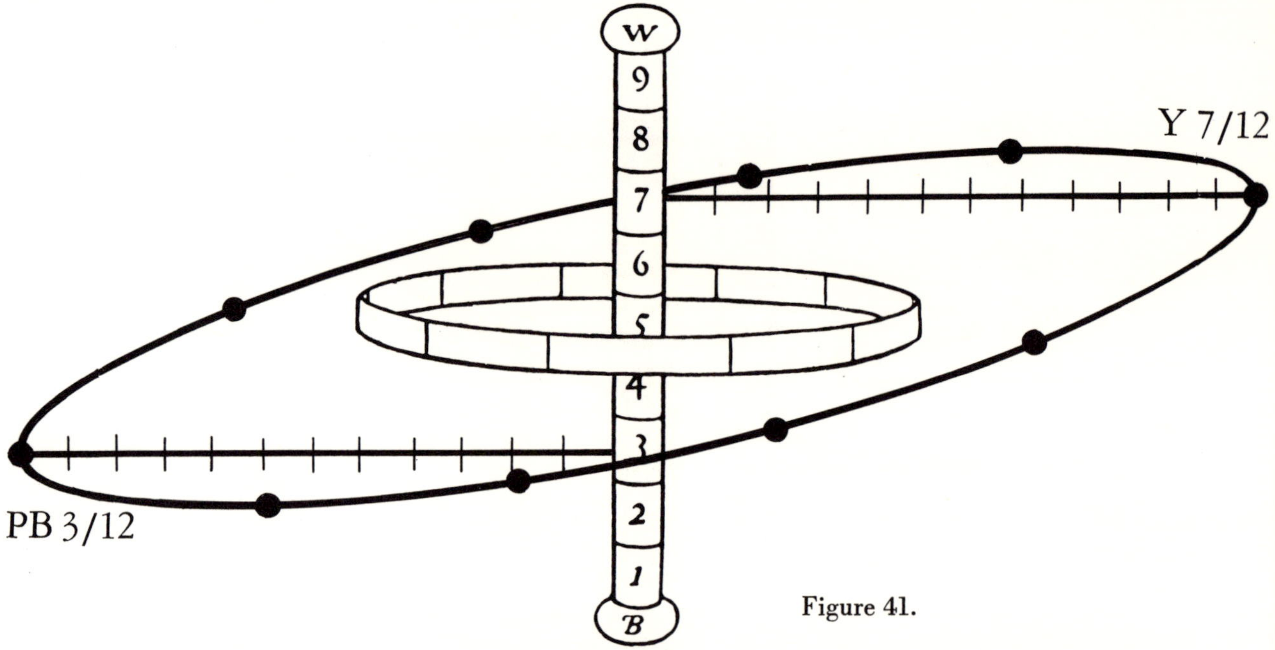

Figure 41.

RP and G, are set at chroma 5, while the other hues in the sequence move in and out toward neutral gray N5: RP 5/5, R 5/4, YR 5/3, Y 5/3, GY 5/4, G 5/5, BG 5/4, B 5/3, PB 5/3, P 5/4, and back to RP 5/5. The same kinds of paths, all at value 5 (or value 6 or 7, or 4 or 3), could be planned.

Figure 41 traces an elliptical path between a high-value yellow and a low-value purple-blue, with Munsell's ten key hues all included in the path. Thus: Y 7/12 scales down to YR 6/8, to R 5/4, to RP 5/4, to P 4/8, to PB 3/12, then up to B 4/8, BG 5/4, G 5/4, YG 6/8 and back to Y 7/12. This harmony is seen in illustration A on Color Plate VIII. The key colors, yellow and purple-blue, are fairly pure in chroma, and they are given beautiful emphasis and richness through the support of the weaker and softer chromas of the hues that lie between them on the elliptical path.

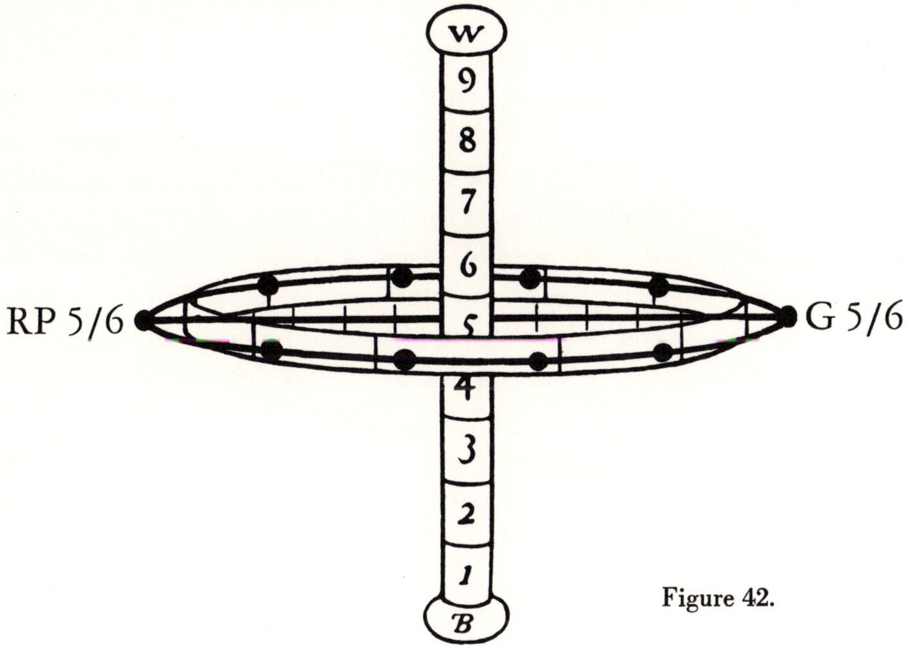

Figure 42.

Elliptical paths can be traced in virtually any direction and angle on the Munsell Color Sphere. In the above example one of the colors, yellow, is fairly pure in chroma and light in value, while the other, purple-blue, is deeper in value and also pure.

Figure 42 and illustration B on Color Plate VIII trace another path, this one on a black ground. Here two fairly pure colors, red-purple (RP) and green (G), have the same medium 5 value. Sequences toward the warmer colors go *up* in value and are weaker in chroma, while sequences toward the cooler hues go *down* in value and also are weaker in chroma, as follows: RP 5/6, R 6/4, YR 7/2, Y 7/2, GY 6/4, G 5/6, BG 4/4, B 3/2, PB 3/2, P 4/4. The result is quite majestic and most dramatically proves Munsell's contention that order assures beauty.

Rigorous control of color, such as described in this chapter, may seem academic and laborious to many creative minds. Discipline is forever tedious. Yet without some measure of it, color expression may be childish, impulsive, and audacious.

A musician needs to know a great deal about the elements of music before he can compose. Then he can forget the rules. The same with color harmony and the Principles of Albert H. Munsell. To experiment with them, to understand them, is to enlarge the capacity of the artist or designer — regardless of his native talents — and to help him achieve new and original dimensions of color. With training back of him he can then break with it and venture on his own.

Munsell wisely wrote, "Any real progress in color education must come not from a blind imitation of past successes, but by a study into the laws which they exemplify."

The Science of Identification

The Munsell System, as adapted to problems of color notation and description, has become the most highly perfected "tool" to be developed anywhere. Thanks to the cooperation of numerous scientists, to the work of scientific committees, to the endeavors of such organizations as the Inter-Society Color Council, the Optical Society of America, the National Bureau of Standards, the System today is pre-eminent.

Albert H. Munsell originally conceived of a color system that could be used in a "rational way to describe color." There was hopeless confusion with names. He decried such vague terms as pea green, evergreen, invisible green. In color description he noted the borrowing of musical terms such as pitch, key, note, tone, chord, symphony. He wrote, "Can we imagine musical notes called lark, canary, cockatoo, crow, cat, dog or mouse because they bear some distant resemblance to the cries of those animals?"

He saw colors in terms of three dimensions. Using Figure 43 he wrote of peeling an orange and dividing it into five parts, one for reds, one each for yellows, greens, blues, and purples: "The fruit is then filled with assorted colors, graded from white to black, according to their *values*, and disposed by their *hues* in five sections."

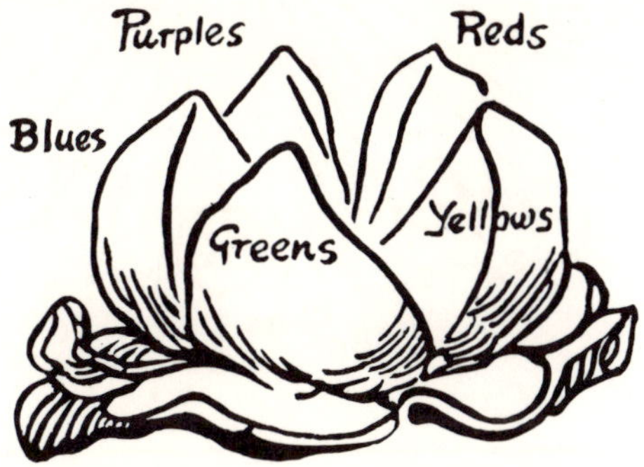

Figure 43.

He then conceived of a color sphere. As Munsell knew, this idea had first been proposed in 1810 by Philipp Otto Runge (1777-1810), a brilliant German painter and contemporary of Goethe. Runge died at the young age of thirty-three. Runge's gradations were to be continuous in all directions. With Munsell, however, he planned a decimal system having five principal hues subdivided into 100 degrees of intermediates, as in Figure 44.

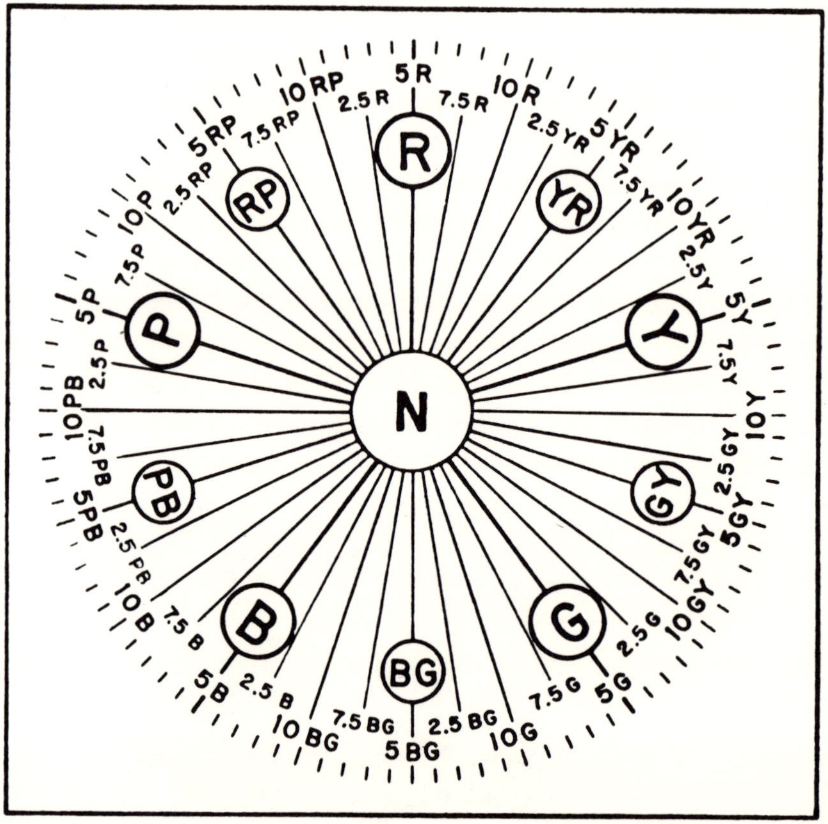

Figure 44.

Key identifications for five principal and five intermediate hues were to be R for red, YR for yellow-red (orange), Y for yellow, GY for green-yellow, G for green, BG for blue-green, B for blue, PB for purple-blue, P for purple, and RP for red-purple. And each of these hues was to have ten degrees or sections on the color circle as its circumference made a complete circuit of the spectrum.

Example: 5R was a principal red; 2.5R was a red inclining toward red purple; 7.5R was a red inclining toward yellow-red, and so on around the circle.

Hue was thus identified by a simple initial or initials like R, RY, RP, or by finer distinctions such as 2.5R, 7.5R, etc. These hues ran about the equator of a sphere.

For *Value* Munsell created a gray scale having O for ideal black, and grading upward in 10 steps to ideal white. To accomplish this he invented a special photometer. (In recent years, his gray scale has been more accurately determined by scientists.) Pure colors had different values or degrees of brightness. Red had a value of 4, yellow a value of 8, green a value of 5, blue a value of 4, and purple a value of 4. The dimension of value was set by comparing the apparent brightness of a color or a modification of color with the brightness of a corresponding step on the gray scale. Value scales in his system ran vertically. Color values were appended to the hue identification by a numeral. R 5/ would be a middle red, R 7/ a light red and R 3/ a dark red.

But his notation was still incomplete, for a third dimension had to be added.

As to the dimension of *Chroma*, Munsell showed remarkable wisdom and ingenuity. He noted, quite truly, that colors and modifications of color, though of the same value could have different degrees of strength or purity. For this dimension he devised the term *Chroma*. Red, for example, was a much "stronger" color than blue-green; it extended further from the gray scale.

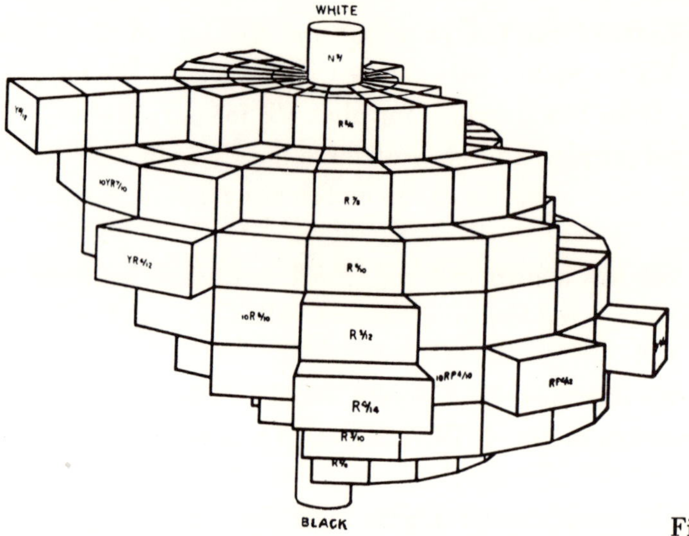

Figure 45.

Munsell, in fact, used blue-green as a sort of anchor, giving it a chroma of 5. Red on the other hand extended outward to a chroma step of 10 or more. (A chroma of 10 was the strongest for red in Munsell's day but has since been extended to more steps.)

Chroma steps and degrees of chroma could readily be established by Munsell through the use of Maxwell disks spun on an electric motor. A fairly pure blue-green, for example, if placed at step 5 (value 5) on the chroma scale would automatically determine a red chroma of 5 when the red was reduced in strength until an *equal* area of it with blue-green would cancel into neutral gray in a Maxwell disk mixture. (Scientists have since designed other and more accurate measuring methods.)

Chroma scales thus ran horizontally out from the gray scale and were given numbers as follows: R 5/ became R 5/2 or R 5/3 if it had a soft rosy tone, or R 5/12 or R 5/14 if it were intense. R 5/14, as a case in point, was a brilliant red with a medium value of 5, and a vivid chroma of 14.

Because chroma steps varied from hue to hue, Munsell's Sphere became a Tree having branches of different length. This is indicated in Figure 45, a modern interpretation of the Munsell Solid. (Also see Cleland's illustration, Figure 11.)

It is quite noticeable in studying Munsell that he placed a neutral gray N5 at the very heart or center of his System and ran his colors and color variations outward from it, but in vertical and horizontal scales. This N5 gray was therefore a vital point of balance, and Munsell favored color schemes which moved in sequence through it and, indeed, cancelled on it.

As to color description, any color, regardless of its hue, value (lightness or darkness), chroma (purity or saturation), can be specified accurately in Munsell notation. Extremely intense or saturated colors, such as red-orange, for example, can be given chroma steps that extend far out from the gray scale. This is a unique feature with Munsell and one that invariably delights the scientific mind.

Here are a few typical Munsell notations. The first identification is hue, the second is value, and the third is chroma. (Wherever the numeral 5 precedes a letter such as R, Y, G, B, P it signifies one of Munsell's key hues. If desired the 5 can be omitted.)

A typical pink: 5R 9/3.
A typical maroon: 2.5R 2/6.
A bright vermilion: 7.5R 5/14.
A rich brown: 5YR 3/4.
A pale ivory: 5Y 9/3.
A grass green: 2.5G 6/8.
A sky blue: 10BG 7/4.
A rich violet: 10PB 3/10.
A lavender: 5P 8/4.

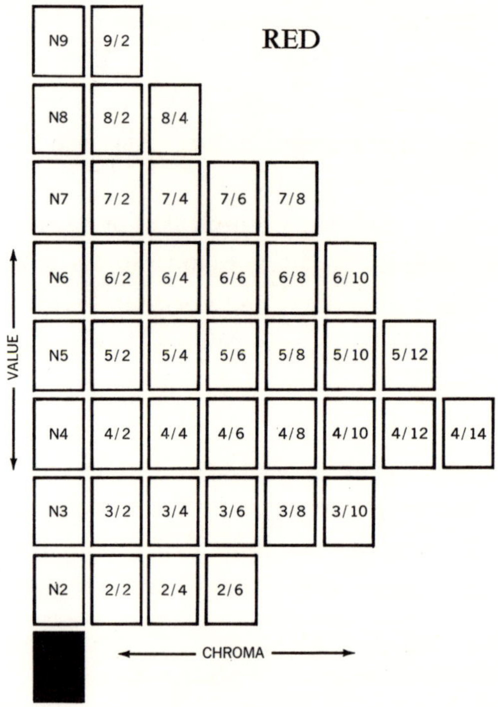

Figure 46.

Yet if Munsell replaced the need for words by substituting numbers, his disciples have recently reversed the process. Look at Figures 46 and 47. There is in existence today a novel method of converting Munsell notations to a descriptive terminology. This is known as *The ISCC-NBS* (Inter-Society Color Council and National Bureau of Standards) *Method of Designating Colors, and a Dictionary of Color Names.* It is known as Circular 553 and is available from the Superintendent of Documents, U.S. Government Printing Office, Washington, D.C. 20402, at a nominal cost of about $2.00. Some 31 hue sections of the Munsell System are charted in black and white with simple names and adjectives that can be used to describe the colors, tints, shades, and

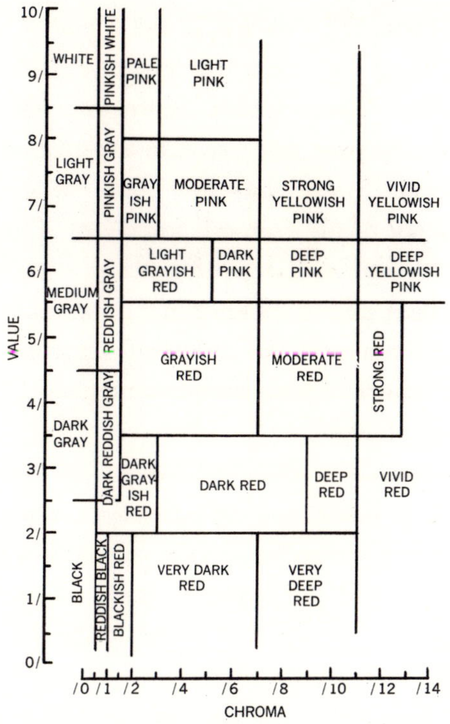

Figure 47.

tones on them. A color chart supplement to this publication is now available from the Office of Standard Reference Materials, NBS, Washington, D.C. 20234 as Standard Sample No. 2106. It is priced at $5.00 per set of 261 Centroid Colors for the ISCC-NBS Color Names.

Figure 46 exemplifies a constant hue red chart (R) from the Munsell System, the original, of course, being in actual color. Figure 47 then suggests names. R 8/4 becomes Light Pink. R 5/4 becomes Grayish Red. R 5/12 becomes Strong Red. R 2/4 becomes Dark Red or Very Dark Red. On around the color circle the same translating can be done for other colors and modifications of color, referring to other charts.

Munsell wanted his system to be useful in the sciences as well as the arts. Though he was an artist in his own right, his greatest posthumous success has been in the field of science. The scientist, more than the artist, has shown greater reverence for him. Anyone interested in the original design and later development of the Munsell System has two excellent sources of reference, both in issues of the *Journal of the Optical Society of America.* The first appeared in December, 1940, and includes reference to a diary kept by Munsell on the creation of his System and to its scientific applications. The second appeared in July, 1943, and includes the report of a special committee on the spacing of the Munsell colors. This special committee report today is the basis of the current *Book of Color.*

But as Munsell continuously sought good technical balance in color, his contributions to esthetics, to color and art education, are today witnessing a renaissance. Satisfied that his system was quite sound as a method of describing color with simple letters and numerals, he wanted this precision and control to apply also to beauty. He wrote, "So must the art of the colorist be furnished with a scientific basis and a clear form of color notation. This will record the successes and failures of the past, and aid in a search, by contrast and analyses, for the fundamentals of color balance. Without a measured and systematic notation, attempts to describe color harmony only produce hazy generalities of little value in describing our sensations."

References

In 1905 *A Color Notation* by Albert H. Munsell was published in Boston. This has since been reissued a number of times with various deletions and changes and has run into eleven editions, any of which would be valuable to consult. The latest edition may be secured from the Munsell Color Company, 2441 North Calvert Street, Baltimore, Maryland 21218.

As to expositions on Munsell's theories of harmony, *A Grammar of Color* by T. M. Cleland, issued by the Strathmore Paper Company in 1921, is reprinted for the most part in the beginning pages and chapters of this book. In 1929 F. G. Cooper wrote the *Munsell Manual of Color*. This was issued by the Munsell Color Company but is now out of print.

Excellent material will be found in *The Art of Color and Design* by Maitland Graves, a sympathetic admirer of Munsell, McGraw-Hill, New York, 1941 and 1952.

On the scientific side, and more related to the Munsell System as a method of color identification, references will be found in *Color in Business, Science and Industry* by Deane B. Judd, John Wiley & Sons, New York, 1952 and 1963; *The Science of Color*, a Report of the Committee on Colorimetry of the Optical Society of America, Thomas Y. Crowell, New York, 1963.

Of particular interest and value are certain teaching aids offered by the Munsell Color Company. These are reasonable in cost and have been widely used in color and art education. Available are a number of charts and chart sets on which color chips may be mounted. There is a hue, value/chroma chart showing Munsell's five principal colors and five intermediate colors, a nine-step gray scale, and a seven-step chroma scale for red. In addition there are sets of constant hue charts for Munsell's key hue (similar to Color Plate II). There is a teacher's demonstration kit and special large wall charts.

The Munsell Book of Color is offered in two collections, one featuring matte finish chips permanently mounted on the charts, the other featuring glossy finish chips which may be readily removed from the charts. The matte collection displays 1,150 color chips. The glossy collection displays 1,450 chips. These publications show the Munsell System in a generous and complete way. They are indispensable for purposes of color identification and notation and helpful in the study of Munsell's principles of color harmony.

COLOR PLATE I

A drawing of the Munsell Color Sphere will be seen in Figure 1, page 10, and a drawing of the Munsell Color Tree in Figure 11, page 26. (Also see Figure 45, page 74.) According to Munsell, color has three dimensions — hue, value, and chroma. On Color Plate I, following page, Munsell's 5 simple hues (red, yellow, green, blue, purple) and 5 compound hues (yellow-red, green-yellow, blue-green, purple-blue, and red-purple) are shown in maximum purity or chroma at upper right and lower left, while the same 10 hues are shown at medium value and medium chroma (5/5) on the other illustrations. The complete Munsell Color Circle has allowance for 100 hues, as charted in Figure 44, page 72.

The first dimension, hue, is traced in a circular direction around a neutral gray axis having 9 value steps. Thus hue is the dimension that distinguishes red from yellow, yellow from green, green from blue, etc.

The dimension of value runs vertically, light colors such as yellow having light value and deep colors such as purple having low value.

The dimension of chroma runs horizontally from the gray scale. Hues of strong chroma, such as red, lie more distant from the neutral gray axis than do hues of weaker chroma, such as blue-green. Chroma steps for yellow at value 5 are included — together with Munsell's 5 simple hues and 5 compound hues at value 5, chroma 5 (5/5). (See Chapter on "The Science of Identification.")

COLOR PLATE I

COLOR PLATE II →

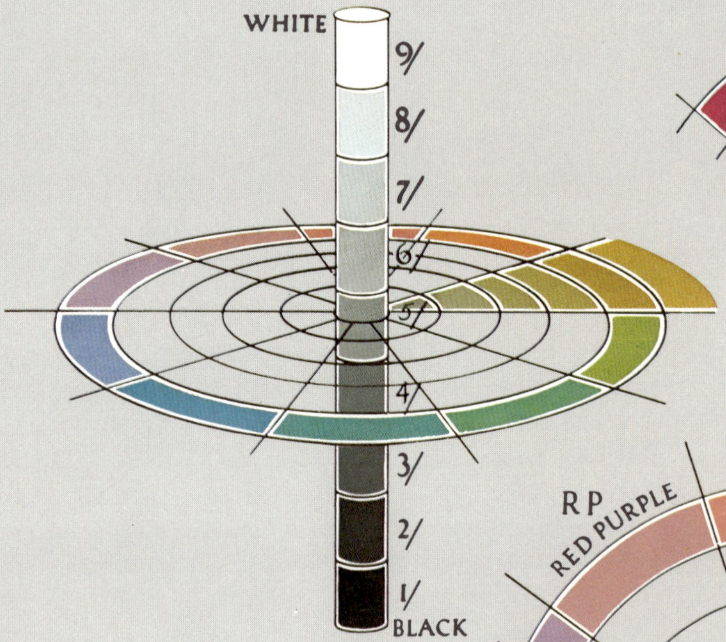

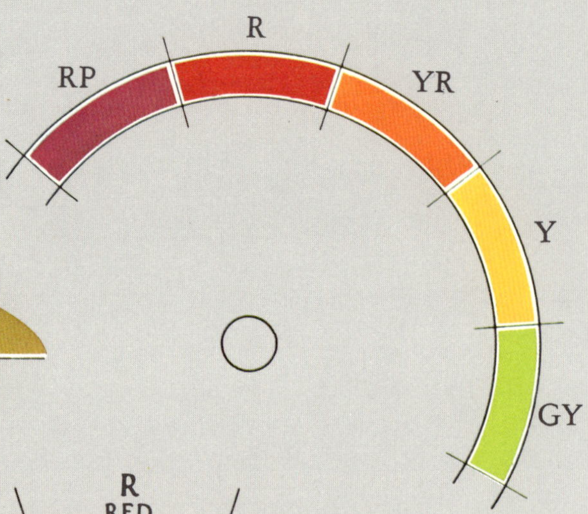

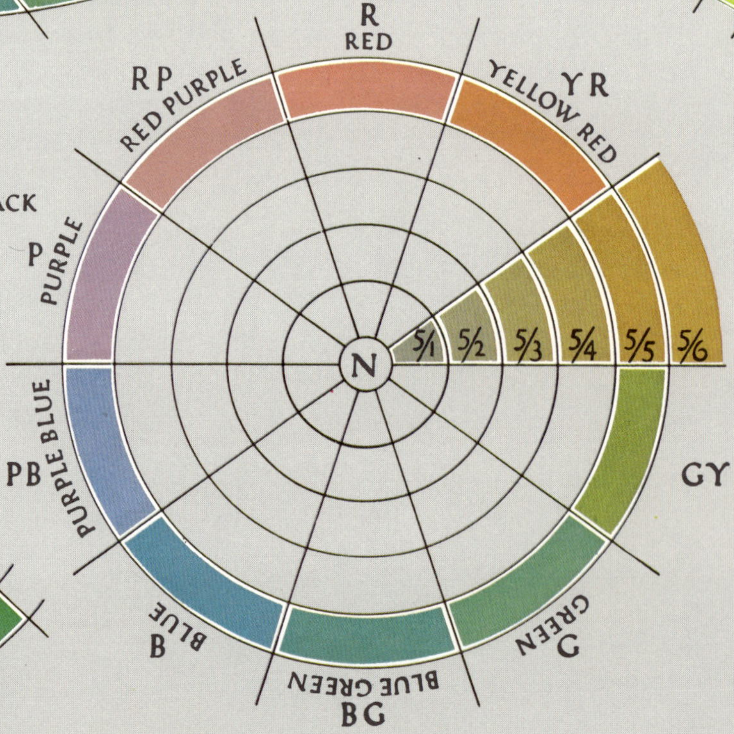

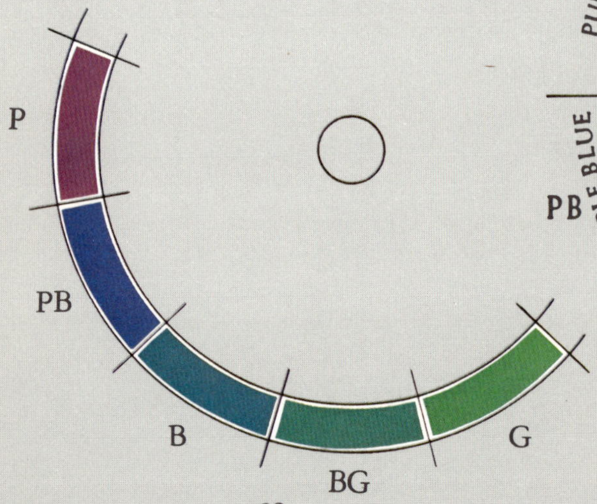

YELLOW

Y

COLOR PLATE II

This illustration (preceding page) shows the yellow (constant hue) chart of the Munsell Color Solid. Note that yellow has its maximum purity or chroma at value 8. At this value there are 12 chroma steps, 7 of which are included on the diagram: 8/1, 8/2, 8/4, 8/6, 8/8, 8/10, 8/12. There are 8 chroma steps for yellow at value 9; 10 at value 7; 8 at value 6; 6 at value 5; 4 at value 4; 2 at value 3; and 1 at value 2.

Thus *vertical* scales for yellow show light and dark variations having the *same chroma. Horizontal* scales show weak and strong variations or chromas for yellow having the *same value.*

Of Munsell's 5 principal colors and 5 compound colors shown on Color Plate I, these reach maximum chroma at different values. Light colors, such as yellow-red and green-yellow, will naturally reach full or maximum chroma at high values, while deep colors, such as blue, purple-blue, and purple, will reach full or maximum chroma at low values. Constant hue charts for these colors will thus take different form than shown for yellow. See Figure 28, page 50, for an example of yellow-orange and Figure 46, page 76, for an example of red.

COLOR PLATE III

The color illustrations on the following page show neatly balanced and related color arrangements of different values and chromas of yellow-red (YR). (Also see Figure 28, page 50, and accompanying discussion in the text.) To Munsell, beauty followed good order. Value steps or chroma steps of any one color (or any group of colors) pleased the eye when sequences were well arranged in vertical, horizontal, or diagonal directions or paths.

In illustration A on Color Plate III the dimension of chroma is held constant at /4, while values step from 3/, to 5/, to 7/, to white. Here all tones have the same apparent chroma or purity.

In illustration B, the dimension of value is held constant at 5/, and the stepping is in a horizontal direction from a fairly pure yellow-red, 5/12, to 5/8, to 5/4, to neutral gray N6. This effect is more subtle.

In illustration C the path for yellow-red is diagonal from a high value tone that is fairly pure (7/6) to a medium tone (5/4) to a deep tone for the background which is near gray (3/2).

In illustration D, the path is again diagonal but opposite to illustration C. Here the light tone used for the background is grayish (8/2) and the harmony swings through a medium tone (6/4) to a fairly rich deep tone (4/8).

COLOR PLATE III

A

YR 3/4, YR 5/4, YR 7/4, White

B

YR 5/12, YR 5/8, YR 5/4, N5

C

YR 7/6, YR 5/4, YR 3/2

D

YR 4/8, YR 6/4, YR 8/2

COLOR PLATE IV

A

R 5/5, BG 5/5, N5

B

R 6/5, BG 4/5, N5

R 5/5

Munsell's 5/5 colors (middle value and medium chroma) have always been admired in the graphic arts and in fine printing. Two loyal exponents of Munsell who both wrote monographs on his theories of harmony, T. M. Cleland and F. G. Cooper, were typographers and type designers. Color in fine printing and typography demands good balance where

Y 5/5

G 5/5

B 5/5

C

P 5/5

COLOR PLATE IV

Munsell had a special liking for colors of medium value and medium chroma (5/5) and examples are shown on Color Plate IV, preceding page. In illustration A, a 5/5 red is combined with a 5/5 complementary blue-green on a neutral N5 ground. If this tends to appear unduly muted, Munsell suggested that one tone might be raised a step in value and one lowered a step. In illustration B, middle gray N5 is maintained, but the red has been raised to 6/5 and the blue-green lowered to 4/5. Note that chroma or purity remains unchanged. This effect will probably be found more pleasing than illustration A.

In C, five medium value, medium chroma colors (all 5/5) are shown for red, yellow, green, blue, purple. Because these colors have true elegance and because they lie visually midway between white and black, they have been the delight of printers and typographers. Where used in design or decoration with black type on white paper, the eye senses that the order is in ideal balance.

COLOR PLATE V

The illustrations on the following page show color combinations of opposite hues having different value and different chroma — but which none the less follow well-related steps and paths within the Munsell color solid. In A, a high value yellow of strong chroma (8/12) is combined with a deep value purple-blue of weak chroma (2/6). The yellow is three steps above medium value N5, and the blue is three steps below.

In B, a similar arrangement of yellow-red and blue is charted, but the middle gray (N5) is used as a background. Here the yellow is two steps above middle value and the blue is two steps below. In this instance the more chromatic yellow-red is given smaller area than the softer blue — a practice which Munsell endorsed.

In C, a light green-yellow of strong chroma is similarly combined with a deep purple of weak chroma, and again on a middle gray ground. The purer green-yellow is given smaller area than the purple.

In D, a clean tint of red-purple 6/8 is combined with a soft tone of green 4/4. It is visually best to use the warmer color (i.e. red, yellow, yellow-red) for the light tone and the cooler color (i.e. green, blue, purple) for the deeper.

COLOR PLATE V

A

Y 8/12, PB 2/6

B

YR 7/10, B 3/5, N5

C

GY 7/10, P 3/5, N5

D

RP 6/8, G 4/4

COLOR PLATE VI

A

P 5/5, RP 7/5, PB 3/5

B

Y 8/12, YR 6/8, GY 4/4

C

R 6/10, G 4/5, B 2/5

D

GY 7/10, RP 5/8, PB 3/6

COLOR PLATE VI

Munsell liked complementary hue arrangements but also spoke favorably of combinations of neighboring hues and split complements. Neighboring hues are those that lie next to each other on the color circle. In two examples on the preceding page, A shows a purple, a red-purple, and a purple-blue (all good neighbors) in which the dimension of chroma is held constant at /5, but in which the red-purple is given a high value 7/ and the purple-blue a deep value 3/. Note that the sequence is through a middle-value, middle-chroma purple 5/5 for neat balance. B shows the sequence of neighboring yellow-red, yellow, and green-yellow which runs diagonally from a high value strong chroma yellow (8/12), to a slightly toned yellow-red (6/8), to a more modified green-yellow (4/4).

In illustration C, a high value red (6/10) is combined in a diagonal path with its split-complements, green 4/5 and blue 2/5. The warm red (pink) is given smaller area.

In illustration D, the split-complements of a fairly pure green-yellow (7/10) are also combined in a diagonal path with deeper and softer red-purple (5/8) and purple-blue (3/6).

COLOR PLATE VII

As advanced principles of color arrangement, Munsell developed and offered what he termed diminishing sequences and elliptical paths. Two striking examples of a diminishing sequence of colors are shown on the following page. (Also see Figure 40, page 67.) In illustration A, the sequence begins with a fairly pure yellow of high value and relatively strong chroma (8/9). It then swings around the cool side of the color solid, *moving down one value and in one chroma* — (toward gray) as it proceeds — to green-yellow 7/8, green 6/7, blue-green 5/6, blue 4/5, to purple-blue 3/4 for the background.

In illustration B, the two end points of the diminishing sequence are the same as in A (Y 8/9 and PB 3/4), but the path swings around the warm side of the color solid — from Y 8/9 to YR 7/8, to R 6/7, to RP 5/6, to P 4/5, to PB 3/4. There is little doubt but that both these color effects have true beauty — as Munsell insisted, because in both cases neat visual order is followed and featured.

A

Y 8/9, GY 7/8, G 6/7, BG 5/6, B 4/5, PB 3/4

B

Y 8/9, YR 7/8, R 6/7, RP 5/6, P 4/5, PB 3/4

COLOR PLATE VIII

A

Y 7/12, YR 6/8, R 5/4, RP 5/4, P 4/8, PB 3/12, B 4/8, BG 5/4, G 5/4, YG 6/8

B

RP 5/6, R 6/4, YR 7/2, Y 7/2, GY 6/4, G 5/6, BG 4/4, B 3/2, PB 3/2, P 4/4

COLOR PLATE VIII

The harmony of the elliptical path represented, to Munsell, one of the most advanced of all principles. Examples are charted in Figures 21, 41, and 42, pages 38, 68, 69. An artist or designer could select a preferred hue (preferably with its complement) and then trace an elliptical path into and around the color solid. As the sequence moved, each step would rise up or down in value, or in and out in chroma (or both), thus presenting a wholly natural and logical order, such order being assured of successful esthetic result.

In illustration A on Color Plate VIII, preceding page, an elliptical path is shown, starting with a pure yellow of strong chroma (7/12). The progression of steps then moves down in value and inward in chroma to YR 6/8, R 5/4, RP 5/4, at which point the value path continues to lead downward, but the chroma steps go outward again toward purity until PB 3/12 is reached (this is the background in illustration A). Now the path goes around, inward and upward on the other side of the solid through B 4/8, BG 5/4, G 5/4, YG 6/8 and back again to the key yellow 7/12.

In illustration B, two complementary key hues of the same value and chroma are located at red-purple 5/6 and green 5/6. Starting with RP 5/6 the elliptical path rises *upward* and inward and then outward on the warm side of the solid to its opposite G 5/6. From here the swing is *downward* and around in similar fashion on the cool side of the solid and back to RP 5/6.